Chinese Mathematics

中國數學

李儼
杜石然
著

郭樹理
倫華祥
譯

CHINESE MATHEMATICS
A concise history

by

Lǐ Yǎn and Dù Shírán

translated by

John N. Crossley
Professor of Mathematics
Monash University

and

Anthony W.-C. Lun
Principal Lecturer in Mathematics
Hong Kong Polytechnic

CLARENDON PRESS • OXFORD
1987

Oxford University Press, Walton Street, Oxford OX2 6DP

Oxford   New York   Toronto
Delhi   Bombay   Calcutta   Madras   Karachi
Petaling Jaya   Singapore   Hong Kong   Tokyo
Nairobi   Dar es Salaam   Cape Town
Melbourne   Auckland

and associated companies in
Beirut   Berlin   Ibadan   Nicosia

Oxford is a trade mark of Oxford University Press

Published in the United States
by Oxford University Press, New York

British Library Cataloguing in Publication Data
Lǐ, Yǎn
Chinese mathematics: a concise history.
1. Mathematics—China—History
2. Mathematics, Chinese—History
I. Title   II. Dà, Shirán   III. Zhong guo shu xue. English
510'.951   QA27.C5
ISBN 0-19-858181-5

Library of Congress Cataloging in Publication Data
Li, Yen, 1892-1963.
Chinese mathematics.
Translation of: Chung-kuo ku tai shu hsüeh chien shih.
Title on added t.p.: Chung-kuo shu hsüeh.
1. Mathematics—China—History.   2. Mathematics, Chinese.
I. Tu, Shih-jan,   II. Crossley, John N.   III. Lun, Anthony Wah-Cheung.
IV. Title.   V. Title: Chung-kuo shu hsüeh.
QA27.C5L4713   1986   510'.951   86-17955
ISBN 0-19-858181-5

Typeset by Colset Private Limited, Singapore
Printed in Great Britain by
The Alden Press, Oxford

Foreword

Sir Joseph Needham, Director of the Science and Civilisation in China Project, The Needham Research Institute, Cambridge

It is a great pleasure and honour for me to have been asked to write a foreword for this translation of a new book on the history of mathematics in ancient China. Tu Shih-Jan of Academia Sinica is a personal friend of mine, but Li Nien, unfortunately, I never met when I was in China during World War II, though I gather that he was one of the officials in the Northwest Highways Administration centred on Lanchow. Nevertheless, I had a boundless admiration for his books on the history of mathematics, the very type of all the best works on the history of science in ancient China. The usual pronunciation of Li's name would be Li Yan, but I remember he was always called Li Nien because that pronunciation was used in the part of the country that he himself came from.

As will be seen from the present book, mathematics had a very great development in ancient China. This was perhaps to be expected in view of the advanced nature of their astronomy. If China developed no Euclidean deductive geometry, there was plenty of empirical geometry there; and if one thinks of the 'crystalline celestial spheres', one might argue that Europe had more geometry than was good for it. In any case, perhaps like the ancient Babylonians, the Chinese always preferred algebraic methods; and indeed by the thirteenth century AD, they were the best algebraists in the world.

Mathematics has a particularly important role in the history of science because the great Scientific Revolution, which brought all modern science, ecumenical and universal, into existence, depended essentially upon the mathematization of hypotheses about Nature, combined with relentless experimentation. Now mathematics is a universal language, which any trained person can understand, and therefore far above all national languages. The Chinese were particularly prepared for this, because the characters of their written script have exactly the same meaning, no matter how you pronounce them—and in some dialects the pronunciation might be very strange.

This is why mathematics matters so much, and it is a great joy to me that the work of Li Nien and his student, Tu Shih-Jan, should have been so well translated into the English language. I hope it will have a very wide circulation, as it deserves to have, so that those who want to know more about ancient Chinese mathematics will not have to learn Chinese and plough through the books on the subject (as I had to do myself). May I wish every good thing for this fascinating book.

Author's Preface

I was very happy to learn that the joint work by Professor Lǐ Yǎn and myself, *Chinese Mathematics: A Concise History*, had been translated into English by Professor J. N. Crossley and Dr Anthony W. -C. Lun of Monash University, Australia and that it is now almost ready for publication by the Oxford University Press.

Chinese mathematics, as introduced in this book, is a completely different system from the Greek mathematics of the West yet this style of mathematics has existed continuously in China for more than two thousand years. Through this book it is possible to get some understanding of Chinese mathematicians and the results of their researches in mathematics throughout these two thousand years. I believe that this sort of understanding will produce considerable benefits for the development of comparative historical research in mathematics.

Professor Lǐ Yǎn (1892–1963) was my supervisor when I was a graduate student. Originally he was a railway engineer but in the first decades of this century he began his work researching in the history of Chinese mathematics as an amateur. This work continued unceasingly for several decades until his death. In the course of his life he published over one hundred books and papers which culminated in his five volumes of *Collected Essays on the History of Chinese Mathematics*. He also produced other work of a more specialized nature. His two volume *Survey of the History of Chinese Mathematics* can be regarded as representative of his work. In 1955 he was transfered to the Academia Sinica in Beijing. I wrote the original of this book under his guidance when he was ill and it was published after his death. It can be regarded as his last work. The present English translation forms a fitting memorial to Professor Lǐ Yǎn.

Professor Crossley and Dr Lun have devoted a great deal of effort into the difficult task of translating the ancient Chinese mathematics into fluent modern English. I am fully aware of how difficult this work is and I wish to extend my deepest gratitude to them.

In order to help the reader to understand the recent developments in research on Chinese mathematics I have added several short passages, dealing with recently excavated materials.

Finally I wish to thank Sir Joseph Needham for his splendid foreword to the English translation.

Beijing   Dù Shírán
June 1986

Translators' Preface

Sir Joseph Needham's *Science and Civilisation in Chinese Society* (1959) Volume 3 Part I gives an excellent introduction to Chinese mathematics for the Western reader. In the present book we present a more detailed Chinese view. In making the translation we have tried to stay as close as possible to the original. The major difficulty we encountered was translating excerpts from ancient Chinese works. Classical Chinese is a study in its own right and we, as mathematicians, acknowledge our inadequacy here.

There are only three significant changes we have made from the original work. First, we have incorporated the footnotes into the text. Second, because we have made the translation for those who are interested in China but hampered by the Chinese script, we have omitted references to articles and books in Chinese and Japanese and added references to articles and books in Western languages whenever we could. The interested reader should also consult the extensive bibliography in Ang and Swetz (1984). Third, we have included in the Appendix very brief introductions (with further references) to the language and history of China. (These were written for mathematicians, not historians or linguists.)

We have followed an ancient practice in translating, dating from at least eleventh century Spain, in having one native speaker of each language. However we have relied on many people for advice and assistance. Chief among these is Professor Dù Shirán of the Institute for the History of Natural Science, Béijing, one of the original authors. He has supplied us with new data, which we have incorporated in the text. Professor Dù has also read the whole translation with great care and the manuscript is much improved thanks to his help. Others who have helped in many ways include: Mr Ng Kan Chuen who first drew our attention to the book in 1980, Dr Lam Lay Yong of the National University of Singapore and Dr D. B. Wagner of Copenhagen, who have given us many helpful criticisms and advice, Dr J. P. Hu of Monash University who gave continual help with the many aspects of the translation, President Li Xiang of Guizhou University and Dr J. C. Stillwell of Monash University, the East Asian History of Science Library, Cambridge, which provided many new Western language articles, and the Hargrave and Main Libraries of Monash University, who helped in tracking down materials in Australia. Finally we thank the many people who have helped with the onerous job of preparing the manuscript: principally Ms Anne-Marie Vandenberg and Miss Mary Beal who each typed the whole manuscript, but also Mrs Irene Bouette, Mrs Nora Fleming, Mrs Edwina Fisher, Mrs Lynn Kemp, Mrs Carol Brown, Mrs Pam Keating, and Mr Kevin Grace. To all of these we are very grateful. Any omissions or errors are solely attributable to us, the translators.

Monash University
September 1985

J. N. C.
A. W.–C. L.

Contents

Contents

1
The beginnings of mathematics in ancient China
(before the Qín Dynasty (before 221 BC))

1.1 The very beginnings of the concepts of number and geometric figure

1 On the origin of numbers, legends of quipu knots, and the gnomon and compass

In order to discern the very beginnings of mathematics in Ancient China we have to trace back into the distant past. If one asks 'When did the people of Ancient China begin to grasp the rudiments of the concepts of number and geometric figure?', it is not easy to answer the question. Just like trying to answer 'Who were the first to use fire, stone axes, or hoes?', we cannot give a precise answer to 'Who was the first to know how to calculate with numbers?'

Because no-one can answer this question precisely, all sorts of legends and myths were created.

The *Shì Běn* (世本, the *Book on Ancestries*) is an ancient book on Prehistoric China. In the *Book on Ancestries* there is recorded a legend: 'The Yellow Emperor [who is alleged to have reigned 2698–2598 BC] (黄帝, Huángdì) ordered his subjects Xī Hé (義和) to observe the sun, Cháng Yí (常儀) to observe the moon, Yú Qū (臾區) to observe the stars, Líng Lún (伶倫) to fix musical scales, Dà Náo (大撓), to determine the methods for fixing time and the seasons and ordered Lì Shǒu (隸首) to create arithmetic'. In ancient times the legend of Lì Shǒu creating arithmetic circulated widely. It is contained not only in the *Book on Ancestries* but also quoted in many other old texts. (The *Book on Ancestries* itself is lost but scattered passages from it are quoted or referred to in other ancient Chinese books. The passage just quoted is in the 'Explanatory Notes' to Sīmǎ Qiān's *Chronicles: Book on the Calendar* by Sīmǎ Zhēn (司馬貞) written in the Tang dynasty about the eighth century AD. (See Chavannes 1895 Vol. III, p. 323 fn. 1 and Vol. I, p. 32 fn. 2.)

To credit the creation of the concept of number to one man, that is, to credit it to Lì Shǒu in the time of the Yellow Emperor, is obviously not in accord with the historical facts. In fact the concept of number cannot be the invention of a single genius. It is only possible that in the long history of mankind the concept of number gradually evolved because of the practical requirements of human activity.

Besides 'Lì Shǒu creating arithmetic', we must mention two other legends

1

on the very beginnings of ancient mathematics. These concern quipu knots and the gnomon and compass. (We use the translation quipu because of the similarity to the Peruvian system of knots.)

The chapter Xìcí in the *Yì Jīng* (易·繫辭, the Great Appendix to the *Book of Changes*) says: 'In prehistoric times men used knots, later on sages replaced these by writing'. Zhuāngzǐ (莊子) also recorded the following legend: 'In the ancient past, during the time of the Róng Chéng (容成), Xuān Yuán (軒轅), Fú Xī (伏羲) and Shén Nóng (神農), people tied knots to communicate.'

According to these quotations, before written language was invented, men used string to tie different types of knots to remind themselves of particular matters. It is not unreasonable to infer that people of ancient times also used knots to record numbers. Some ancient books (of a later period) went further, explaining: 'For a major matter, use string to tie a big knot; for a minor matter, tie a small knot. The number of knots corresponds to the number of matters'. According to various sources, until not so long ago there were still a few stone-age tribes using quipu knots to record numbers.

'Guī (規)' means compass, the instrument for drawing circles; the shape of 'Jǔ (矩)' is similar to the present-day set-square (gnomon) used by carpenters, an instrument for drawing rectangles. 'Guījǔ (規矩)' is a phrase nowadays used frequently in China to mean rules and regulations but its use can be traced back into the past. These are instruments also produced in the mythical period.

According to legend Chuí (倕) invented the gnomon and compass. Some ancient books say Chuí invented the gnomon and compass and the plumb-line, and then the whole country followed him in the use of compass, carpenter's square, and plumb-line. However, in other legends it is claimed

Fig. 1.1 The gnomon and compass. (From the ancestral temple of the Wǔ family of the Hàn dynasty.)

that Fú Xī (伏義), the first of the Three Sovereigns, was the inventor of Guijǔ. On a stone relief of the Han Dynasty, about the second century AD, which has been preserved to the present day, Fú Xī (伏義) is pictured holding a gnomon and Nǔ Wā (女媧) holding a compass. There are other similar stone reliefs. According to these legends the gnomon and compass probably appeared very early. In the *Chronicles* of Sǐmǎ Qiān it is also mentioned that while the Emperor Yǔ (禹) of Xià (reigned *ca.* 21st century BC) was attending to floods in the country, he went about 'with a plumb-line in his left hand and a gnomon and compass in his right' while carrying out survey work in the course of bringing the floods under control. In fact in Sǐmǎ Qiān's *Chronicles*, Book 2, the History of the Xià, the original text says: 'Through the four seasons [Yǔ] rode on chariots on land, rode in boats on water, skied on mud and in hilly places rode in a sedan chair, in his left hand he held a plumb-line, in his right a gnomon and compass, in order to rid the land of floods and open up the nine districts and the nine roads [i.e. China].'

Although most of these legends were created by people of later periods imagining the real situation, we can see from these myths that:

(a) from the very early stages, in those ancient times which cannot accurately be described, people began to grasp concepts relating to number and geometric figure;

(b) before recorded history, quipu knots were used to record events and numbers;

(c) from the very early stages, instruments like the gnomon and compass were used to draw simple patterns.

2 Decimal notation

In exploring how ancient mathematics evolved, it is, of course, possible to infer certain events from legends and myths. However, more important clues are found in ancient cultural artifacts that have been excavated. More accurate deductions can then be made through archaeological investigations.

According to evidence from artifacts revealed by excavation, about a hundred thousand years ago the 'Hé Tào (河套) Man' inscribed diamond-shaped patterns on bone utensils. By that period stone implements had already taken on certain definite shapes. In the more advanced 'Yǎngsháo (仰韶) Culture' (about 5000 BC), in addition to animal patterns on earthenware there were some geometric patterns. Some of these patterns were formed from combinations of triangles and straight lines while others were formed from circular dots and curved lines. There were also criss-cross and chequered patterns. It was also discovered that on some of this earthenware there were special signs and marks that had been purposely inscribed on the implements by these ancient people. Most of these marks were vertical lines; some were Z signs.

After many tens of thousands of years of primitive civilization, a society with a class structure was formed (about 2000 BC). This was the slave society of the Shāng dynasty (*c.* 16th–11th centuries BC).

It is clear from archaeological discoveries that the culture of the Shang dynasty was fairly well developed. Because of further advances in agriculture, a division of labour was fostered in society. Rectangular and circular grain storage buildings belonging to slave owners of those times have been excavated at Zhèngzhōu (鄭州), Huìxiàn (輝縣), and other places. The use of bronze had also developed further. There were long, short and other shapes of weapons made of bronze. Rectangular and circular bronze utensils for eating and vessels for sacrificial purposes were also manufactured. Following further developments in the division of labour in society the exchange of products expanded. Coins with holes in the middle, which were used in those days, have also been discovered at Zhèngzhōu.

About the 14th century BC the Shāng Dynasty moved its capital to the neighbourhood of the present-day Xiǎotún (小屯) near Ānyáng (安陽) in Hénán province. Culture and the economy took a further step forward.

In the later stages of the Shāng Dynasty some form of calendar appeared because of the requirements of agricultural development.

Since the end of the nineteenth century (AD) a large collection of plastrons (under-shells) of the land-tortoise and animal bones with characters inscribed have been excavated at Xiǎotún near Ānyáng in Hénán province. Research has shown that the nobles of the Shāng period worshipped the spirits of their ancestors. They prayed to their ancestors' spirits and asked all kinds of questions. The questions asked, the answers, and sometimes the subsequent verifications are inscribed on the plastrons and animal bones.

The type of Chinese characters in the inscriptions on the plastrons and bones of the Shāng period is generally known as 'oracle bone script'. In this oracle bone script the phrases and sentences are those of divination. (See for example Chang Kwang-chih 1980, pp. 31–42.)

'Oracle bone script' is the earliest form of Chinese writing so far discovered, though isolated symbols have been found on Yǎngsháo pottery. It can be said that there are written records of the history of China from that period. The oracle bone scripts are valuable materials for understanding the late period of the Shāng Dynasty.

From the excavated oracle bones we know that the Shāng people used up to five thousand characters. Amongst these are numerals. These are the earliest historical records of numerals in China. Often on these oracle bones there are recorded the numbers of prisoners taken in a war or the number of the enemy killed, the number of birds and animals caught while hunting, the number of domestic animals killed for sacrifices to spirits, etc. On the oracle bones numbers of days counted were also recorded. For example: 'The eighth day, namely Xīnhài (辛亥), two thousand six hundred fifty six men were killed while crossing spears.'

'Captured ten and six men.'
'Ten dogs and five dogs.'
'Ten cattle and five.'
'Deer fifty and six.'
'Five hundreds four tens and seven days.'

The ordinal number Xīnhài in the first quotation is represented in a sexagesimal system. (See below, p. 22.) The largest number on the oracle bones is 30 000, the smallest is one. Units, tens, thousands and ten thousands each have a special character to represent them.

Their ideograms from one to ten in oracle bone script are roughly as follows:

Oracle:	一	二	三	亖	X	∩、∧	+	)(	九	|
Modern:	一	二	三	四	五	六	七	八	九	十
Hindu:	1	2	3	4	5	6	7	8	9	10

Hundred, thousand and ten thousand have the forms:

Oracle:			
Modern:	百	千	萬
	hundred	thousand	ten thousand

Twenty, thirty, . . ., two hundred, three hundred, . . . and two thousand, thirty thousand, etc. are compound words made up of two characters, for example

二十	三十	四十	五十	六十	七十	八十
20	30	40	50	60	70	80
二百	三百	四百	五百	六百	八百	九百
200	300	400	500	600	800	900
二千	三千	四千	五千	八千	三萬	
2000	3000	4000	5000	8000	30 000	

For instance, 'two thousand six hundred and fifty-six' would be written in oracle bone script as […].

Among ancient scripts that have been preserved up to the present, in addition to the oracle bone script there is a script inscribed on bronze. This kind of script is known as the 'bell vessel script' or 'bronze script'. Research shows that most of the bronze script was the written language in the Zhōu period (*ca.* 11th century–221 BC). In the bronze script, numerals were written in a similar way to those on the oracle bones. The only differences are that ten is written as 'ϕ' and four, besides being written '≡', is also written as '四' or '四'. In the bronze script compound words are written differently from on the oracle bones. Thus 'six hundred and fifty-nine' in bronze script is written as […] . Here, '又' (modern 又, yòu, meaning 'and') is used to separate units, tens, hundreds, thousands, etc., thus giving 'six hundred and fifty and nine'. Moreover, fifty is written as a compound word with five on top and ten below. This is just the opposite to the oracle bone script, where five is at the bottom and ten on top.

In the Hàn Dynasty (206 BC–220 AD) the character '又' used for separation in recording (large) numbers was dropped and compound numerals also disappeared. The shape of character then used is almost exactly the same as the present day Chinese.

In Table 1.1 below we have written the characters for one to ten in oracle bone script, in bronze script, in the script used in the Hàn Dynasty, in present-day Chinese characters and in the modern Western, called Hindu-Arabic numerals, so that it can be seen how the characters evolved.

Table 1.1

Oracle bone	一	二	三	亖	X	∩, ∧	+	)(	九	|
Bronze	一	二	三	三, 亖	三, X	介	+	)l	九	♦
Hàn	一	二	三	四	X	六	七	八	九	十
Present Chinese	一	二	三	四	五	六	七	八	九	十
Modern Western	1	2	3	4	5	6	7	8	9	10

1.2 Calculation with counting rods — the principal calculating method in ancient China

1 The origins of calculating with counting rods

In the previous section decimal notation was discussed. However, in ancient times calculations did not directly involve manipulating numerals. The device used for calculations by the ancient Chinese was counting rods.

In the history of the evolution of human civilization different races have used different devices for calculation. We know that about four to five thousand years ago the Babylonians (who lived in present-day Iraq) used wedge-shaped sticks to press into clay tablets. They recorded numerals by making cuneiform signs and they performed calculations. The ancient Egyptians used hieroglyphs. They recorded many types of mathematical problems on papyrus made from the reed which grows along the Nile. They also performed calculations on papyrus. In the Middle Ages, Hindus and Arabs performed calculations on sand boards or directly on the floor using small pointed sticks. 'Counting rods' are a device peculiar to ancient China.

The counting rods, called 'Chóu' (籌), are small bamboo rods. Ancient Chinese mathematicians operated with these short bamboo rods by arranging them into different configurations to represent numbers and then performed calculations using these rods. Manipulating these 'chóu' to perform calcula-tions is known as 'chóu suàn' (calculating with chóu 籌算).

In the *Analytical Dictionary of Characters* (説文解字, *Shuōwén jiězi*) by

Xŭ Shèn (許慎) of the Eastern Hàn Dynasty (120 AD) who is usually regarded as the father of Chinese lexicography, there is one character 筭 (suàn), and another character (算, suàn). Xŭ Shèn's explanation of the character is as follows: '筭, six [Chinese] inches long, a tool for calculating calendars and numbers. Composed of "zhú" 竹 [as radical: see appendix] meaning bamboo and "nòng" 弄 meaning to play with something, thus indicating that by frequently manipulating these rods there will be no mistake in calculations.' Xŭ Shèn's explanation of the character 算 is: '算', meaning to calculate. Composed of bamboo radical 竹 (on top), with 具 jù, meaning tool or device, below. Read as 'suàn', exactly like the other character 筭. Scholars in the Qīng Dynasty (1644–1911 AD) explained the difference between the two characters 'suàn' thus: ' 筭 is a device for calculation, therefore the character is a noun; while 算 means "to calculate with bamboo rods" and is therefore a verb'.

In the Lǜ Lì Zhì chapter of *Qián Hàn Shū* ('Memoir on the calendar' in *History of the Western Hàn Dynasty* 前漢書 · 律曆志, first century AD) is recorded: 'To calculate one uses bamboo of diameter one 'fēn' [分, a tenth of an inch], six [Chinese] inches long [about 140 mm].' In the Lǜ Lì Zhì chapter of the *Suí Shū* (隋書 · 律曆志, 'Memoir on the calendar' chapter of *History of the Suí Dynasty*, seventh century AD) there is also reported: 'To calculate one uses bamboo, two 'fēn' [a fifth of an inch] wide, three inches long [about 70 mm]'. (In future we shall use 'inch' to mean 'Chinese inch'. The lengths of such measures were different at different times in ancient China, varying between about 22.5 and 33.3 mm.)

From these passages we can infer that from the Hàn Dynasty to the Suí (隋) Dynasty the counting rods used were gradually shortened. This is probably because smaller counting rods are easier to manipulate for calculations.

Up to now no reliable evidence has been found to determine when counting rods started to be used for calculations. However, it is plausible that by no later than the Warring States period (481–221 BC), people were quite familiar with the techniques of using counting rods to perform calculations. Texts concerning that period which have survived to the present used the ideograms 'chóu' (籌) and 'suàn' (筭). For instance in the *Lǎozǐ* (老子; *The Book of Master Lǎo*) there is a statement: 'Those well-versed in calculation use neither counting rods nor texts'. That is to say, such people knew how to calculate without using counting rods, they calculated in their heads. Their are several chapters in the *Book of Rites* (儀禮; *Yí Lǐ*) that mention the ideogram suàn (筭). For example, in the chapter, 'Rite for offering special sacrifice' (特牲饋食禮; Té shēng kuì shì lì) 'the wine vessels are beyond counting', where suàn 筭 (counting rod) is used to mean 'counting'. The two chapters 'The rite of archery in the township' (鄉射禮, Xiāng shè lì) and 'Grand archery' (大射, Dà shè) have descriptions of using suàn (筭) to count. And in 'Grand archery' there are descriptions illustrating to some extent the methods for counting in tens and those for counting figures less than ten.

In August 1971 more than 30 rods dating back to the time of Emperor Xuān Dì (宣帝, 73–49 BC) of the Western Hàn Dynasty were found in Qiānyáng County (千陽縣) in Shǎnxī (陝西) province. The length and width of these rods conform to the descriptions in the *History of the Hàn Dynasty*, with the difference that the Shǎnxī rods are made of bone. A bundle of rods were unearthed in 1975 in Han tomb No. 168 at Fènghuāngshān (鳳凰山) in Jiānglíng (江陵) in Húběi (湖北) province. These are made of bamboo, but are a little longer than the Shǎnxī rods. They date from the reign of the Emperor Wén Dì (文帝, 179–157 BC). In 1978 a quantity of earthenware with the signs and marks of rods dating from the time of the Warring States period (475–221 BC) of Eastern Zhōu was found in Dēngfēng County (登封縣) in Hénán (河南) province. All these are unearthed artifacts of counting rods (see Fig. 1.2).

2   Decimal place-value notation in calculations with counting rods

We can now explain how counting rods are used for calculation. There are two ways of using counting rods to represent numbers. The first is the vertical form, the other is the horizontal form. They are shown below:

Hindu form:	1	2	3	4	5	6	7	8	9
Vertical form:	I	II	III	IIII	IIIII	T	TT	TTT	TTTT
Horizontal form:	—	=	≡	≣	≣	⊥	⊥	⊥	⊥

These forms are arranged to accord with a decimal place-value system. Such a system, like the present Western systems, involves counting in multiples of tens, hundreds, thousands, etc. so that ten units make a ten, ten tens make a hundred, ten hundreds make a thousand, ten thousands make a ten thousand (萬, read 'wàn', since there is a single Chinese word for 10 000), etc. This is called the 'base ten' system. Digits in a number can indicate units, tens, hundreds, thousands, ten thousands and so on by virtue of their position. For example, in 17, 74, and 6 708 the digit 7 indicates seven, seventy, and seven hundred, respectively. This is known as a 'place-value' system. Nowadays the standard method of writing numbers is a place-value system to base ten. Hence it is called a decimal place-value system for recording numbers. For recording angles one uses $1° = 60'$, $1' = 60''$ (one degree equals 60 minutes, one minute equals 60 seconds). This is a sexagesimal (base sixty) system, not a base ten system. The Romans wrote XXII to represent twenty-two where X means ten and I is one. Although this is a 'base ten' system, the positions of the symbols do not have particular meanings. Thus it is not a place-value system.

In the oracle bone script in China units, tens, hundreds, thousands, and ten thousands are represented by special ideograms and they are in base ten. That means to say that the base ten system was used very early on in China.

(a)

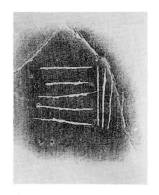

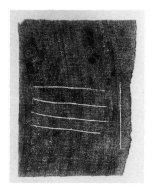

(b)

Fig. 1.2 (a) The Shǎnxī rods. (b) The signs of rods on earthenware excavated in Dēngfēng County in Hénán province.

Recording numbers by counting rods in ancient China is in base ten and is a place-value notation. This is clearly a decimal place-value system for writing numbers.

How are the counting rod forms arranged to accord with the decimal number system? The method is: for the units use the vertical form, for the tens use the horizontal form, for the hundreds use the vertical form, for the thousands use the horizontal form, for the ten thousands again use the vertical form and so on. A blank space is used for zero. Hence a number can be represented by digits arranged in vertical and horizontal forms alternatively, working from right to left in the usual order units, tens, hundreds, thousands, ten thousands, etc. For example, 378 can be represented as ⫼ ⊥ 𝍠 , 6 708 as ⊥𝍢 𝍠 . This method of recording numbers is explained in *Master Sūn's Mathematical Manual* (孫子算經 ; *Sūnzǐ Suànjīng*, about the fifth century AD) and in *Xiàhóu Yáng's Mathematical Manual* (夏侯陽算經 ; *Xiàhóu Yáng Suànjīng*, about the eighth century AD). Master Sūn says:

'Units are vertical, tens are horizontal,
Hundreds stand, thousands lie down;
Thus thousands and tens look the same,
Ten thousands and hundreds look alike'.

In the original text of *Xiàhóu Yáng's Mathematical Manual* he says:

'Units stand vertical, tens are horizontal,
Hundreds stand, thousands lie down.
Thousands and tens look the same,
Ten thousands and hundreds look alike.
Once bigger than six,
Five is on top;
Six does not accumulate,
Five does not stand alone'.

The last four lines say that for digits equal to or larger than six, one of ⊥⊥ ⊥ ⊥ or 𝍢 𝍣𝍤𝍥 must be used to represent them. The counting rod on top (in either form) is used to stand for five rods. This is just the same as for the Chinese abacus, where each bead above the crossbar stands for five beads below. When we get to the digit six, this number is not represented by piling up counting rods, thus: not ☰ or ⦀⦀ but ⏤ or 𝍢 . 'Five does not stand alone' means that the digit five must not itself follow the 'one for five' method described above, which uses one counting rod to represent five; this is so as not to be confused with the representation in the tens, thousands, etc. positions.

Written language in ancient China was arranged in columns reading from top to bottom and right to left. However, when recording numbers using counting rods, the characters are arranged from left to right and therefore this is the same as the present way of recording numbers in both East and West.

Although in the history of mathematics many countries of the world adopted the base ten system from the very beginning, the use of a decimal place-value system came much later. In the West, the reading and writing of numbers in the past was quite complicated. For example, the ancient Greeks used twenty-four standard Greek letters and three ancient letters to represent numbers less than a thousand. So the methods of calculation had to be complicated. The decimal number system we use nowadays (commonly known as Arabic or Hindu-Arabic numerals) has its origins in India. The Hindu decimal number system appearing on artifacts excavated to date cannot be older than the 6th century AD.

The counting rod method of recording numbers used by the ancient Chinese has simple and convenient characteristics. Take 7 + 8 as an example, i.e. add ⊤ and ⫪ . At one glance the two counting rods on top go together to make ten, while the rods below add up to five. This easily gives 7 + 8 = 15. It is obvious that this method of addition, using counting rods, is easier to master than the commonly used calculation techniques with Arabic numerals. This is because it can be understood easily just by looking at it.

In ancient China, the decimal place-value method of recording numbers appeared at the latest during the Spring and Autumn period (770–476 BC) or the Warring States period (475–221 BC). Hence from the Spring and Autumn or Warring States periods onwards various arithmetic operations could be performed easily and conveniently.

The decimal place-value counting rod notation was a notable invention in the mathematics of ancient China. It was not until the middle of the fifteenth century AD that the use of the abacus became very common. Counting rods were an important device for calculations in ancient Chinese mathematics. Calculating by means of counting rods is the key to understanding the mathematics of ancient China. It is also a distinctive feature of ancient Chinese mathematics.

3 Arithmetic operations with counting rods

The four basic arithmetic operations were in use very early in China as the following example, recorded in a book on law and order, *A Treatise on Law* (法經, *Fǎjīng*) edited by Lǐ Kuī (李悝) during the Warring States period shows. (To understand the example one needs to know that a 'mǔ' is a measure of area, approximately 0.067 ha, a 'dàn' is a volume of approximately 1 hl and that bronze coins were used as far back as the eighth century BC.)

A farmer with a family of five has a hundred mǔ (畝) under cultivation. Each year each mǔ yields a harvest of one and a half dàn (石), thus giving one hundred and fifty dàn of cereal. After the tax of fifteen dàn, one tenth of the total, there remains one hundred and thirty-five. Each month, each member of the family consumes one and a half dàn; five persons consume a total of ninety dàn of cereal each year. There remain forty-five dàn. Each dàn is worth thirty coins, a total of one thousand three hundred and fifty. Three hundred coins are used for ceremonial and sacrifices at the ancestors' shrine. Thus there are a thousand and fifty left. Each person spends three hundred coins on clothing, so five people spend one thousand five hundred each year. Consequently there is a deficit of four hundred and fifty.

In this passage there are addition, subtraction, and multiplication. One could also say that division was involved.

One presumes that the calculations for addition and subtraction were invented earlier than those for multiplication and division. However, it is difficult to give a definite answer as to when these arithmetic operations began. In the ancient texts that have survived to the present there is no explicit mention of the calculations for addition and subtraction. This is because addition and subtraction are very simple so they do not require detailed discussion. On the other hand, one cannot avoid using addition and subtraction in multiplication and division.

From ancient mathematical manuals that recorded methods of multiplication and division (and therefore involved addition and subtraction) and from additions and subtractions on the abacus (recall that we mentioned above that the abacus evolved from calculations using counting rods) we can deduce the following:

1. Additions and subtractions using counting rods were performed like multiplications and divisions, that is, starting from the highest place-value digit and therefore calculating from left to right. This is just the opposite of the usual presently adopted method of calculation. In modern calculations one starts from the lowest place-value digit, thus calculating from right to left.

2. This method of calculation is essentially the same as for additions and subtractions on abaci. The only difference is that beads are replaced by counting rods. For abacus calculations there are also addition and subtraction mnemonics, but similar mnemonics for additions and subtractions using counting rods have not yet been found.

Example: 456 + 789 using counting rods. First use counting rods to represent 456, then add 7 to the 4 in the hundreds' position. Second, add the numbers in the tens' and then in the units' position. So one starts from the highest place-value digit, calculating from left to right as follows:

$$+ \begin{cases} 7 & 8 & 9 \\ 4 & 5 & 6 \end{cases} + \begin{cases} 7 & & \\ 4 & 5 & 6 \end{cases} + \begin{cases} & 8 & \\ 1 & 1 & 5 & 6 \end{cases} + \begin{cases} & & 9 \\ 1 & 2 & 3 & 6 \end{cases}$$

                            1 1 5 6   1 2 3 6   1 2 4 5

Subtraction is similar. For instance, 1234 − 789. First lay down 1245 and subtract 7 from the hundreds' position. Second, subtract the numbers from the tens' and then the units' positions. Again this is carried out from left to right as below:

$$-\begin{cases} 1\ 2\ 4\ 5 \\ \ \ \ \ 7\ 8\ 9 \end{cases} -\begin{cases} 1\ 2\ 4\ 5 \\ \ \ \ \ 7 \end{cases} -\begin{cases} 5\ 4\ 5 \\ \ \ 8 \end{cases} -\begin{cases} 4\ 6\ 5 \\ \ \ \ \ 9 \end{cases}$$

− Ⅱ ≡ ⅢⅠ	ⅢⅠ ≡ ⅢⅠ	Ⅲ ⊥ ⅢⅠ	Ⅲ ≡ Ⅰ

| | 5 4 5 | 4 6 5 | <u>4 5 6</u> |

Ⅱ ≟ ⅢⅠ

As far as multiplication and division are concerned, every Chinese knows they are connected with the 'Nine-nines rhyme' for multiplication. 'One one makes one, one two makes two, one three makes three, . . . one nine makes nine. Two twos make four, two threes make six, . . .'. Nowadays every Chinese primary school pupil knows that by heart. But why is this rhyme called 'Nine-nines'? When did this 'Nine-nines' start?

In ancient times this rhyme was different from the present one which all Chinese school children know by heart. It started with 'Nine nines make eighty-one'. So it was called the 'Nine-nines rhyme'. Consequently the name 'Nine-nines rhyme' was adopted and is still used today. It seems from the available information that the 'Nine-nines rhyme' started with 'Nine nines are eighty-one' and went down to 'two twos are four', and remained thus until the first or second century AD. This is evident from Hàn bamboo strips that have been excavated. The extension of the Nine-nines rhyme down to 'One one is one' occurred somewhere between the fifth and tenth centuries AD. The rhyme recorded in *Master Sūn's Mathematical Manual* (孫子算經) is of this form. In the thirteenth or fourteenth century AD, during the time of the Sòng Dynasty, the order of the Nine-nines rhyme was reversed and changed to that presently used, that is, starting from 'One one is one' going to 'Nine nines are eighty-one'.

As to when the ancient Chinese started using the 'Nine-nines rhyme' there is a story that goes like this. During the Spring and Autumn period (770–476 BC) Duke Huàn (桓公) of the feudal state of Qí (齊) established an 'Institute for recruiting distinguished persons' (招賢館, Zhāo xián guǎn) in order to attract people with outstanding abilities. Although he waited quite a while there were still no people applying for positions. After a year a man came along and brought with him a 'Nine-nines rhyme' as a gift for the Duke and to show his outstanding knowledge. Duke Huán thought it was quite a joke and said to the man: 'You think the Nine-nines rhyme demonstrates some kind of advanced knowledge?' The man gave him a sound answer. He said: 'As a matter of fact, knowing the Nine-nines rhyme is not sufficient to show any

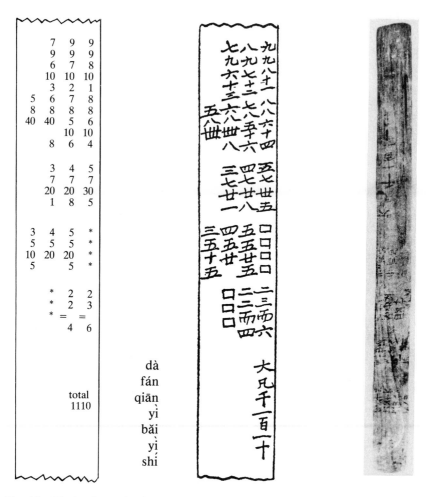

	7	9	9
	9	9	9
	6	7	8
	10	10	10
	3	2	1
5	6	7	8
8	8	8	8
40	40	5	6
		10	10
	8	6	4
	3	4	5
	7	7	7
	20	20	30
	1	8	5
3	4	5	*
5	5	5	*
10	20	20	*
5		5	*
	*	2	2
	*	2	3
	*	=	=
		4	6

total
1110

dà
fán
qiān
yì
bǎi
yì
shí

Fig. 1.3 Hàn bamboo strip with Nine-nines rhyme (Read from top down.)

ability or learning. But, if you appoint me who only know the Nine-nines rhyme then there is no doubt that people of ability and skill will queue up for employment.' Duke Huàn thought this was a sound argument. So he accepted him and warmly welcomed him. In less than a month many people of ability and skill applied, coming from many different places.

This story implies that by no later than the Spring and Autumn and Warring States periods the Nine-nines rhyme had been adopted pretty widely in China. In fact it was quite a common thing at that time. In many texts of that period that have survived to the present, such as *the Book of Master Xún* (荀子, *Xúnzǐ*), the *Book of Master Guǎn*, (管子, *Guǎnzǐ*) etc. there are many events recorded connected with the Nine-nines rhyme and many passages containing lines from the Nine-nines rhyme.

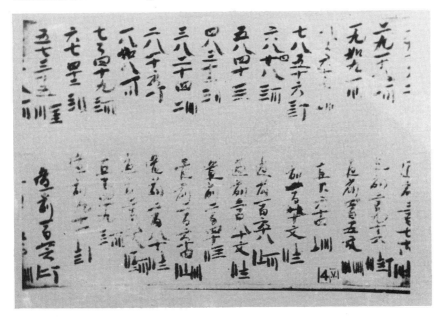

Fig. 1.4 The Dūnhuáng scroll: section of a nine-nines rhyme in 'A Short Course on Calculation Techniques' (立成算法 Lì chèng suàn fǎ) of the Táng Dynasty. (The original is in the British Museum (S. 930). The end of each line is in counting rod configurations such as ⚏ (81), ⊥‖ (72), etc. This is the earliest material concerning calculation with counting rods.)

Since the end of the 19th century AD many bamboo and wood strips with traces of Chinese characters have frequently been excavated in North-West China (see Fig. 1.3). Most of these bamboo and wood strips date from the Hàn Dynasty (206 BC–220 AD). They are usually known as 'Hàn strips'. Some of these strips record the Nine-nines rhyme. However, these bamboo and wood strips had been buried in the ground for a long time and the Nine-nines rhymes on them are not complete. There remain on them only three or four lines in general. One strip contains fourteen lines and the most found has been seventeen lines. These are the oldest Nine-nines rhymes visible in China today. In the Dūnhuáng scroll (敦煌, Fig.1.4) there is a Nine-nines rhyme from the Táng Dynasty (618–907 AD).

The method of using counting rods for multiplication was described in great detail in *Master Sūn's Mathematical Manual* and *Xiàhóu Yáng's Mathematical Manual* (夏侯陽算經, *Xiàhóu Yáng suànjīng*). The two numbers in a multiplication are arranged one above the other, so that the highest place-value digit of one of the numbers is directly above the lowest place-value digit of the other and there is a row left blank in between these numbers. Next the highest place-value digit on the top row is multiplied by the digits on the bottom row, in turn, from left to right. The products are added up at once after each multiplication and then the result is put into the space between the two numbers. Now move the number in the bottom row one space to the right (so that its lowest place-value digit is now directly under the second highest place-value digit of the top row). Next multiply the second highest place-value digit in the top row by the digits in the bottom row and add the result into the middle row as above. Continue until finally the number obtained in the middle row is the product of the two numbers.

For example, for 234×456 the two numbers are arranged as in diagram (1). The top row is the multiplier, the bottom row is the multiplicand and the middle row is left blank for writing in the product. Then move the multiplicand 456 to the left so that its lowest place-value digit is directly below the highest place-value digit of the multiplier 234. Use the highest place-value digit 2 in 234 to multiply the digits in 456 accordingly. Adding immediately after each multiplication we obtain 912, which is recorded in the middle row, as in diagram (2). Now remove the digit 2 in 234 (to indicate it has been used to multiply) and then move 456 one space to the right as in diagram (3). The second step uses the digit 3 in 234 to multiply the digits of 456 in turn. Adding the result after each multiplication to 912 we obtain 10488. Then also remove the digit 3 in 234 (to indicate it has been used to multiply) and move 456 again one place to the right as in diagram (4). Use the last digit, 4, of 234 to multiply 456. Adding to the number in the middle row after each multiplication we get 106704. Finally remove the digit 4 in 234 to complete the whole multiplication process as in diagram (5).

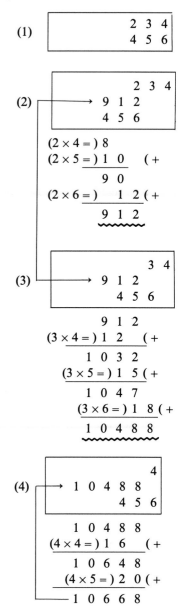

(1)

```
        2 3 4
        4 5 6
```

(2)

```
              2 3 4
    ──────→ 9 1 2
              4 5 6
```

$(2 \times 4 =) 8$
$(2 \times 5 =) 1 \ 0$ $(+$
          $9 \ 0$
$(2 \times 6 =) \quad 1 \ 2 (+$
          $9 \ 1 \ 2$

(3)

```
              3 4
    ──────→ 9 1 2
            4 5 6
```

          $9 \ 1 \ 2$
$(3 \times 4 =) 1 \ 2$ $(+$
        $1 \ 0 \ 3 \ 2$
$(3 \times 5 =) 1 \ 5 (+$
        $1 \ 0 \ 4 \ 7$
$(3 \times 6 =) 1 \ 8 (+$
      $1 \ 0 \ 4 \ 8 \ 8$

(4)

```
                    4
    ──→ 1 0 4 8 8
              4 5 6
```

      $1 \ 0 \ 4 \ 8 \ 8$
$(4 \times 4 =) 1 \ 6$ $(+$
      $1 \ 0 \ 6 \ 4 \ 8$
$(4 \times 5 =) 2 \ 0 (+$
    $1 \ 0 \ 6 \ 6 \ 8$
$(4 \times 6 =) 2 \ 4 (+$
    $1 \ 0 \ 6 \ 7 \ 0 \ 4$

(5)

```
      1 0 6 7 0 4
            4 5 6
```

In the above example we used Arabic numerals for the step-by-step procedure of multiplication by counting rods shown in the diagrams to make it easy for the present-day reader to understand. In fact, where calculating using counting rods was done it was natural to record the numbers in counting rod form. So diagram (2) should really be

‖ ≡ ‖‖‖	Top row
⅏ — ‖	Middle row
‖‖‖ ≣ T	Bottom row

The rest can be inferred. The calculations listed below each box of the diagram are, in fact, the steps involving mental calculations and manipulations of chóu — the counting rods. The procedure for division using counting rods is similar, as shown below.

The method for division is the reverse of the method for multiplication. In divisions the dividendum is called 'shi' (實), the divisor is called 'fǎ'(法) and the quotient is called 'shāng' (商). As an example of the method for division using counting rods take 106704 ÷ 456. First arrange the dividendum 106704 above the divisor 456 as in diagram (a). Notice that it is the same as the last diagram (5) in the example above for multiplication. Then we move 456 to the left to the point where 1067 is divided by 456. The quotient, 2, is recorded in the top row, which had been left blank. Using the quotient, 2, multiply each digit of 456 in turn. Subtracting the results after each multiplication from 106704 we obtain the first remainder 15504 as in diagram (b). The second step: move the divisor one space to the right; divide again and we obtain the second digit 3 in the quotient. Multiply each digit of 456 by 3. Subtracting the result after each multiplication from the first remainder 15504 we get the second remainder 1824 as in diagram (c). Finally, move 456 one space to the right again; then divide and we obtain the third digit, 4, of the quotient. Multiply each digit in 456 by 4. Subtracting

(a)
```
        1  0  6  7  0  4
                 4  5  6
```

(b) →
```
                 2
        1  0  6  7  0  4
                 4  5  6
```
```
           1  0  6  7
(2 × 4 =) 8            ( –
           2  6  7
(2 × 5 =) 1  0        ( –
           1  6  7
(2 × 6 =) 1  2        ( –
           1  5  5
```

(c) →
```
                 2  3
           1  5  5  0  4
                 4  5  6
```
```
           1  5  5  0
(3 × 4 =) 1  2        ( –
           3  5  0
(3 × 5 =) 1  5        ( –
           2  0  0
(3 × 6 =) 1  8        ( –
           1  8  2
```

the result after each multiplication from the second remainder 1824 we find the numbers conveniently divide exactly as in diagram (d). There being no remainder indicates that the division process is complete and the quotient is 234 as in diagram (e).

The method of division using counting rods is very similar to the 'Galley Method' (which was still widely used in the West in the sixteenth century AD) that was introduced into China from the West towards the end of the Ming Dynasty (1368–1644 AD). However, in divisions using counting rods, subtraction takes place after each multiplication, so the counting rods are constantly being re-arranged. In the Galley method each step is recorded by pen. Using 106704 ÷ 456 as an example again, the Galley method is as follows in diagrammatic form:

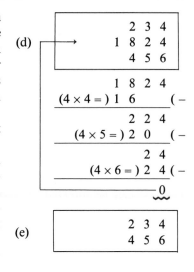

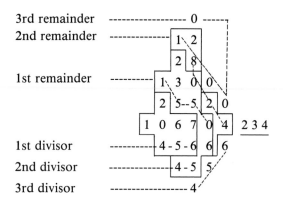

If the two numbers do not divide exactly, the final arrangement of the counting rods is such that the top row is the integral part of the quotient, the middle row, which is in the position of the dividendum, represents the numerator and the bottom row, which is in the position of the divisor, represents the denominator of the fractional part of the quotient. For instance

$$106838 \div 456 = 234 \frac{134}{456}$$

The corresponding diagram of the counting rods would be:

that is

2	3	4	shāng
1	3	4	shì
4	5	6	fǎ

1.3 Mathematical knowledge in ancient texts before the Qín Dynasty (221–206 BC) and mathematical education

1   Mathematical knowledge in the *Book of Crafts* (考工記, *Kǎo Gōng Jì*), the *Book of Master Mò* (墨經, *Mòzǐ*) and other ancient texts

In the various writings of scholars in the period before the Qín Dynasty, besides the books *Xúnzǐ, Guǎnzǐ,* other texts on the Nine-nines rhyme and also the example given in the *Treatise on Law* (法經) by Li Kui (李悝), one can still find some items about mathematical knowledge in other texts. In particular the *Book of Crafts* in the *Record of Rites of the Zhōu Dynasty*, the *Book of Master Mò*, and certain propositions given by logicians such as Hui Shi (惠施) and others, contain passages concerning mathematics.

The *Book of Crafts* in the *Record of Rites of the Zhōu Dynasty* was written, according to research reports, by scholars in the feudal state of Qi (齊) in the Warring States period (475–221 BC). The book was basically written about some construction techniques (such as making horse-drawn carriages and boats and manufacturing bows and arrows) and prescribing various guidelines and uniform dimensions. So it contains some items about fractions, information about angles and the standard measurements and dimensions for containers, etc.

In the chapter 'For craftsmen making the covers for (horse-drawn) carriages' there is a section with the line 'a tenth of a (Chinese) inch is a méi (枚)'. [One méi is what later ages called one fēn (分)]. A lot of books of later periods adopted the method of recording numbers as 'one tenth of an inch', 'two tenths of an inch', and this became the commonest method of recording and writing fractions in ancient China.

The *Book of Crafts* also contains some units used in measuring the size of angles, such as jǔ (矩), xuān (宣), zhú (欘), kē (柯), qingzhé (磬折), etc. Here:

jǔ   = 90°;
xuān   = 45° (= $\frac{1}{2}$ ju);
zhú   = 67° 30′ (= $1\frac{1}{2}$ xuān);
kē   = 101° 15′ (= $1\frac{1}{2}$ zhú);
qingzhé = 151° 52′ 30″ (= $1\frac{1}{2}$ kē).

In the *Book of Crafts* there is also a method for using an arc of a circle to measure the size of an angle. For example, in a section of the chapter 'For craftsmen making bows' there is a passage

'To make bows for the Son of Heaven [the Emperor],
nine bows together make a circle;
To make bows for the feudal lords, seven bows
together make a circle;
Bows for the officials [of the Emperor], five bows
together make a circle;
Bows for the scholars, three bows together make a circle.'

These are the cases where an arc of a circle is used to measure the angle of curvature of a bow. In the astronomy of ancient China the full circle was divided in $365\frac{1}{4}°$ and for an observer on earth the sun moves 1° each day. This is again an example of using a distance on the circumference of a circle to measure the size of an angle.

In general one can say that the concept of angle was no longer considered so important in the development of mathematics in China after the Zhōu (*c.* 11th century–221 BC) and Qín (221–206 BC) dynasties.

In the *Book of Crafts* the standard size of containers is also noted, such as the fú (鬴). The rules are that the fú should be one Chinese foot deep and that the internal size should be one square foot, while the exterior is circular so that the volume is one fú (i.e. 1 cubic foot). In the same place the standard volumes for dòu (豆) and shēng (升) are given. Naturally, the *Book of Crafts* only applied inside the feudal state of Qi. After the unification of China by the first Qín Emperor, there were uniform measurements for lengths, volumes, and weights. In the manufacture of these containers and in finding out their volumes one cannot avoid using mathematical calculations.

Amongst the books written by scholars before the Qín Dynasty the mathematical knowledge contained in the *Book of Master Mò* (墨經) is worth mentioning. The *Book of Master Mò* usually refers to four chapters of *Mòzi*: Book I (經上), Book II, (經下) Explanatory Book I (經説上), and Explanatory Book II (經説下). It is generally accepted that these four chapters were edited by the disciples of Master Mò (墨子). The work begins with a collection of separate statements describing various concepts and definitions. A lot of these statements are connected with logic, mathematics, and physics.

For example, there are some statements about axioms for the basic concepts of geometry in the *Book of Master Mò*, for example:

'Flat: same height,
Straight: three points collinear,
Same length: match up exactly,
Centre: point of same length,
Circle: one centre with same length.'

The description of a circle given here as 'one centre with same length' approaches the modern mathematical definition of a circle as 'the locus of a point which moves at a fixed [i.e. the same] distance from a fixed point [the centre]'. The *Book of Master Mò* also contains lines on the notions of point, line, surface, solid, and the whole solid and its parts (e.g. the hemisphere), and also on the notions of addition and subtraction.

Towards the end of the Warring States period there were also some scholars who abstracted from the physical world and particularized from certain terms and from these they started various debates. These were the logicians (名家) or debaters (辯者) such as Hui Shi (惠施), Gōngsūn Lóng (公孫龍), etc. Some of their paradoxes were, for example, 'eggs have feathers', 'fire is not hot', 'a chicken has three legs', 'a white horse is not a horse', etc. And of course most of these phrases are meaningless and futile. However, there are a few which do reflect some of the mathematical thought of the time. For instance, Gōngsūn Lóng and his co-authors made the following pronouncement:

'A one foot-long stick, though half of it is taken away each day, cannot be exhausted in ten thousand generations.'

This statement is equivalent to the formula

$$\frac{1}{2} + \frac{1}{2^2} + \frac{1}{2^3} + \ldots + \frac{1}{2^n} \ldots \rightarrow 1$$

That is, the total of the sum must approach closer and closer to 1 but it can never be equal to 1. So this pronouncement illustrates a mathematical concept, that is: a finite segment of a line (one foot long) can be represented as the sum of infinitely many finite segments. Later on the method of dividing the circle used by Liú Huī to calculate π was inspired by the above idea. The statement made by Gōngsūn Lóng and others is quite similar to a paradox considered by the Greek philosopher Zeno (in the fifth century BC). The version used by Zeno is as follows. Assume we want to walk from point A to point B. In order to get to B we have first to get to B_1, the midpoint of AB; in order to get to B_1 we first have to reach B_2, the midpoint of AB_1. Continuing in this way there are an infinite number of such midpoints so we can never even start. In the version here it uses a finite segment AB and converts it into an infinite sum of segments (half, then half again) and thence produces an apparent absurdity.

A lot of mathematical textbooks in China nowadays frequently use the example of 'taking half daily' as an example to explain the notion of limit.

2 Mathematical education and the appearance of Sīhuì (司會), Fǎsuàn (法算), and Chóurén (疇人) officials

Besides the writings of the scholars of the Hundred Schools, which contain some comments on mathematics, there are records from those days which also contain reports on mathematical education and about the officials responsible for the statistical calculations, astronomy, and calendarial computations that reflect the development of mathematics in the Spring and Autumn and Warring States periods.

All Chinese know what are called the 'six gentlemanly arts' of the Zhōu dynasty: ritual (禮, lǐ), music (樂 yuè), archery (射 shè), horsemanship (御 yù), calligraphy (書, shū), and mathematics (數, shù). These were the six subjects in the education of the children of the nobility at that time. The *Zhōu Rites* (周禮 *Zhōu lǐ*) records, in particular, the ranks and duties of the officials of the Zhōu Dynasty. In this work there is an entry:

The Bǎo Shì official (保氏) is to teach the ways to the children of the country, to teach the six gentlemanly arts, the first of which is called the five rites, (五禮, wǔ lǐ) the second is the six modes of music (六樂, liǔ yuè), the third the five methods of archery (五射, wǔ shè) the fourth the five ways of horsemanship (五馭, wǔ yù) the fifth the six ways of calligraphy (六書, liǔ shū) and the sixth the nine calculations of mathematics (九數, jiǔ shù).

That is to say, there are officials called the Bǎo Shì (保氏) amongst the various officials and they are responsible for the education of pupils (the children of the country). They are to teach the students the rites, to learn music and archery and how to drive chariots; and mathematics is also one of their subjects.

In another chapter about the system of government in the Zhōu Dynasty in the *Book of Rites*, there are statements such as the following:

'Teach the six-year-olds numbers and directions, . . . the nine-year-olds how to work out days and dates. The ten-year-olds study with a teacher and live away from home learning history, writing, and mathematics.'

This is a description of a curriculum for schoolchildren. The six-year-olds learn 'numbers (數 , shù)', that is, to count from one to ten, 'directions (方名, fāng míng)', that is to recognize north, east, south, and west. The nine-year-olds learn to work out days and dates (數日 , shù rì), that is, they learn to use the Heavenly stems and Earthly branches counting method. The 10 year-olds learn history, writing and calculations, and here calculating (書計, shū jì) means developing mathematical skills. The Heavenly stems and Earthly branches form a sexagesimal cycle for recording. There are ten heavenly stems (甲, 乙 , . . .), and twelve earthly branches (子 , 丑 , . . .). These are taken in pairs, so the first of the cycle is 甲子, the second is 乙丑 , etc. Because there are 10 heavenly stems and 12 earthly branches the scheme repeats after sixty pairs (since 60 is the least common multiple of 10 and 12). The lunar calendar used nowadays in China still uses this system to record the year date. Thus 1986 is the 丙寅 year and 1987 the 丁卯 year.

According to records in the *Zhōu Rites*, at that time there were already officials who were responsible for the census within the boundaries of the Empire. Those officials were called Sīhuaì (司會). Book one of the *Zhōu Rites* says:

'The Sīhuaì consist in total of two senior officials of the middle division, four senior officials from the lower division, eight senior scholars, sixteen middle level scholars, and there are also four archivists, eight secretaries, five petty officials and fifty trainees. Of these, the archivists are responsible for keeping the records and the secretaries, the petty officials and the trainees are responsible for all the errands and mundane jobs.'

From this we can see that this was no small organization. According to Book II of the *Zhōu Rites* the original text recorded that the Sīhuaì were in charge of the accounts of the properties and expenditures of the civil service, the countryside and the local government, and in charge of everything in documents and records (including the accounts, the household records, and maps) and all the duplicates so that they could audit all the monies connected with the officials. So from this we can see that the Sīhuaì were officials responsible for compiling statistics.

In the army there were specialized officials responsible for keeping military statistics, according to the strategy book called the *Liù Tāo* (六韜 , *Book on the Six Arts of War*). It contains a record of a lot of events in the Warring States period. In those days there was a certain rank in the army called 'Fǎsuàn' (法算), two people who were in charge of the weaponry, food and wages for the soldiers, and all the income and expenditure.

In the Hàn Dynasty accountancy had become one specialty among the qualifications of the officials. In the Jū Yán strips (居延　漢簡) recently discovered at Jǔ Yán there are a lot of lines giving the experience and working ability of the officials, which use the words for knowing writing and accounting (能書會計 , néngshǔ huìji), understanding the workings of law and order (通曉律令 , tǒng xiǎo lü lìng), and similar phrases.

Besides the Sīhuaì officials of those days there were special officials involved in astronomy and computing the calendar. Obviously these people had to be well versed in calculation. For example, in the *Zhōu Rites* there is a note about an official called the Féng Xiāng Shì (馮相氏), who was in charge of computing the calendar to determine the four seasons, and another official, the Bǎo Zhāng Shì (保章氏), who was in charge of star charts and determining the movement of the sun, moon, and stars. The following is recorded in the famous *Chronicles* of Sīmǎ Qiān:

During the reign of the Emperors Yōu [幽 , deposed 841 BC] and Li [厲 , killed in 771 BC] [in the Eastern Zhōu Dynasty] the court of Zhōu had declined. The secretaries did not record the seasons and the Emperor ceased informing the people about the dates, so the students of the Chóurén (疇人) were dispersed into the feudal states or into the barbarian states. (See Chavannes 1895 Vol. III, p. 326.)

Undoubtedly these students of the Chóurén (the hereditary officials in charge of astronomy and calendar computing), who were well versed in astronomy and calendar making, must have known mathematics. Much later Ruǎn Yuàn (阮元) (1764–1848 AD) and others of the Qīng Dynasty produced an edition and commentary on the biographies of astronomers and mathematicians of China through the ages. The book is entitled the *Biographies of Chóurén* (疇人傳 , Chóurén zhuàn, written between 1799 and 1898 AD). The meaning of the two characters Chóurén (疇人) has its origin in a sentence in the *Chronicles* of Sīmǎ Qiān.

2
The formation of mathematical systems in ancient China
(Hàn Dynasty, 206 BC–220 AD)

2.1 The Arithmetical Classic of the Gnomon and the Circular Paths of Heaven (周髀算經 , Zhōubì suànjīng)

1 A brief introduction to the contents

In 221 BC the first Qín Emperor united China. Shortly after that the Hàn Dynasty replaced the Qín Dynasty and there was a great increase in productive capacity. This improvement in productive capacity was followed by rapid developments in various areas of science and technology, which in turn fostered the development of mathematics. For example, agricultural production required more accurate forecasts of the seasons and of course this naturally required the ancient people to conduct research into the making of calendars and the study of astronomy. Now the computing of calendars and the study of astronomy require computations and so mathematics, too, developed.

The *Zhōubì suànjīng* (周髀算經) is the earliest Chinese writing on mathematics to have survived to the present day (see Fig. 2.1) and is also a book on astronomy. This book is the result of a gradual accumulation of scientific results from the requirements of astronomy from the Zhōu and Qín periods. The first character in the title is the same as the name of the Zhōu dynasty (周), but it also means perimeter. Bì (髀) is said in the text to mean the gnomon or upright of a sundial. Needham (1959, p. 19) therefore gave the above translation of the title, which we are using, especially since the book is concerned with astronomy and also there are ancient (Sòng Dynasty) Chinese authorities supporting Needham's interpretation of the characters.

A French translation was given by Biot (1841) and part has been translated into Italian by Vacca (1905).

Describing the various schools in the study of the heavens at the end of the Eastern Hàn dynasty, Cài Yōng (蔡邕) (178 AD) said: 'There are three schools in the description of the heavens. The first school is Zhōubì (周髀), the second is Xuān Yè (宣夜) and the third school is Hún Tiān (渾天). (See the *Continuation of the Hàn History* section on astronomy (續漢書 · 天文志, Xù Hàn shū. Tiānwén zhì).) The Hún Tiān school is represented by Zhāng Héng (張衡 , 78–139 AD) in the Hàn dynasty, whose definitive work is called the *Spiritual Constitution of the Universe* (Líng Xiàn, 靈憲). The *Zhōubì* is the representative work on the Gài Tiān (蓋天

25

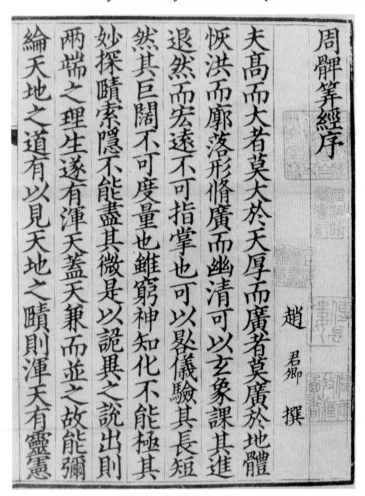

Fig. 2.1 The beginning of the *Zhōubì suànjīng* (from the Southern Sòng edition, now in the Shànghài public library).

covering heaven) theory. The Hún Tiān theory (渾天説) is a more progressive theory which appeared later. The Gài Tiān theory preceded it. (Short descriptions of these three theories may be found in Needham 1959 Vol. 3, pp. 210–24.)

The Gài Tiān theory says that 'the heavens are like an umbrella-hat (蓋笠, gài lì), the earth is like an upturned basin.'

In those days observers of the heavens used to erect a vertical stake and called it a shadow gauge (表, Biǎo). Observing the length of the sun's shadow on the gauge provides data for various calculations. They also used the shadow-gauge for surveying purposes, for example, the illustration given in the *Hǎidǎo suànjīng* (海島算經 , the *Sea Island Mathematical Manual* — see the section on this book, below p. 75).

In the first volume of the *Zhōubǐ suànjīng* it is recorded that people of the Zhōu erected shadow gauges and observed the sun in the capital city of Zhōu and that is why the whole book is called Zhōubǐ (Zhōu shadow gauge). 'Biǎo' simply means shadow gauge; the two characters 算經 (suànjīng, mathematical manual) were added by people of later times (in the Táng Dynasty).

The edition of the *Zhōubǐ suànjīng* which still exists records, near the beginning, a dialogue between the Duke of Zhōu (周公, Zhōu Gōng) and Shāng Gāo (商高), so a lot of people believe this work is from the Zhōu dynasty. But in fact this is incorrect.

First, in the first volume of the *Zhōubǐ suànjīng* there is a passage: 'Mr. Lǔ (呂) said 'Within the four seas, from east to west is 28 000 [Chinese] miles and from north to south is 26 000 miles.' (Under the Gài Tiān theory the world was bounded by four seas and 'within the four seas' has been used as a synonym for China. Like the Chinese inch, the Chinese mile has varied but is about half a kilometre.) The 'Mr. Lǔ' mentioned here was prime minister of the Qín towards the end of the Warring States period, Lǔ Bùwéi (呂不韋). Now in the chapter Yǒu Shǐ Lǎn (有始覽) in *Mr. Lǔ's Spring and Autumn* (呂氏春秋 , *Lǔ shì chūn qiū*), which was edited by Mr. Liǔ's guest scholars, we can find the above sentence. The people of the Zhōu would not have been able to foretell what Mr. Lǔ would have said in the future!

Second, in the *Zhōubǐ suànjīng* many arguments are similar to those in the astronomy manual in the book of *Huái Nán Zǐ* (淮南子) and this book of Huái Nán Zǐ was edited by guests of Liú Ān (劉安), Prince Huái Nán, in the Western Hàn Dynasty. So the book must have been written about the end of the second century BC (that is, in the middle of the Western Hàn Dynasty). Hence we can see that the theories in the *Zhōubǐ suànjīng* were popular during the Qín and Hàn dynasties. Moreover, in the records in the chapter on Arts and Crafts in the *History of the Hàn* (漢書 , Hàn shū) the *Zhōubǐ* book is not listed. It is not until we get to Yáng Xióng (楊雄.) and Cài Yōng (蔡邕) and others in the Hàn Dynasty that the *Zhōubǐ* is mentioned. So we can conclude that, roughly speaking, the edition of the *Zhōubǐ* that is extant was formed into a book between 100 BC and 100 AD (the end of the Western Hàn to the beginning of the Eastern Hàn Dynasty). But it is very probable that some of the contents of this book were known long before the book was written. The earliest surviving edition of the *Zhōubǐ suànjīng* is from the Southern Sòng Dynasty (slightly later than 1213 AD) and is now stored in the Shànghǎi library. There is also an edition published by Zhào Kāiměi (趙開美) towards the end of the Míng Dynasty (1366–1644 AD) and the edition by Dài Zhén ('戴震) from the beginning of the Qīng Dynasty (1644–1911 AD). Present day editions are based on all three of these.

The edition of the *Zhōubǐ suànjīng* that survives has been commented on by many mathematicians of the Hàn and Táng periods. The *Zhōubì suànjīng*, which has been around from the first century BC to the present, has a history of 2000 years. This is an extremely valuable source for our understanding of ancient Chinese mathematics and astronomy. As far as mathematics is concerned, this part of the book expounds the methods of calculation and of surveying heavenly bodies using the Gōugǔ (Pythagoras') theorem, and it tells of complicated calculations with fractions, etc.

2   The dialogue on mathematics between Róng Fāng (榮方) and Master Chén (陳子, Chénzǐ).

The second part of the *Zhōubì suànjīng* records a dialogue between Róng Fāng and Master Chén. At the very beginning, before Master Chén lectures on the Gài Tiān theory, there is a passage on Master Chén's view of the objectives of mathematics and its methods, and his remarks on attitudes towards learning mathematics. The statements in these passages are not only interesting but also have pedagogical value.

Róng Fāng asked Master Chén: 'It is said that . . . the height of the sun, how far its rays can travel and the movement of the sun . . . and the reasons behind these can be found out from what you have learnt, is that true?'

Master Chén replied: 'Yes.'

Róng Fāng asked: 'A fellow like me, can I learn these things?'

Master Chén: 'Of course you can. What you have learnt about the methods of calculation is sufficient for performing all the calculations. What is required beyond this is that you must think earnestly.'

Róng Fāng thought for a few days but still could not grasp the principal idea, so he came again for further instruction. Master Chén said to him in explanation:

That is because you are not very familiar with your own thought. The method of calculation in the study of the heavens is a matter of 'determining the height and measuring the distance', but you still have not got things clear, that is to say, you still cannot generalize what you have learnt. The method of calculation is very simple to explain, but it is of wide application. This is because 'man has a wisdom of analogy' that is to say, after understanding a particular line of argument one can infer various kinds of similar reasoning, or in other words, by asking one question one can reach ten thousand things. When one can draw inferences about other cases from one instance and one is able to generalize, then one can say that one really knows how to calculate. The method of calculation is therefore a sort of wisdom in learning . . . The method of learning: after you have learnt something, beware that what you have learnt is not wide and after you have learnt widely, beware that you have not specialized enough. After specializing you should worry lest you do not have the ability to generalize. So by having people learn similar things and observe similar situations one can find out who is intelligent and who is not. To be able to deduce and then to generalize, that is the mark of an intelligent man . . . If you cannot generalize you have not learnt well enough . . .

In this discourse by Master Chén he first says that mathematics must have a wide application; second, from the abstract point of view of mathematics he is emphasizing the importance of training in deductive and inductive thought. This point is very important for people in understanding and learning mathematics.

3 The Gōugǔ theorem and its use in surveying

The so-called Pythagoras or Gōugǔ (勾股) theorem states that 'in a right-angled triangle, the sum of the squares of the two sides adjacent to the right angle is equal to the square of the hypotenuse.' As illustrated in Fig. 2.2, if a, b are the two adjacent sides and c is the hypotenuse, then

$$a^2 + b^2 = c^2,$$

This is the Gōugǔ theorem. Here the vertical gauge b is called 'Gǔ (股)', meaning a measure or scale. The shadow of the gauge cast by the sun on a plain is called 'Gōu (勾)' and the hypotenuse is known as 'Xián (弦)', the string of a bow. So the right-angled triangle is called 'Gōugǔ shape' (勾股形 , Gōugǔ xíng). From this the theorem of Pythagoras becomes

$$\text{Gōu}^2 + \text{Gǔ}^2 = \text{Xián}^2 \,(\,勾^2 + 股^2 = 弦^2\,).$$

This is a very important theorem and according to legend, when the ancient Greek mathematician Pythagoras (about sixth century BC) established the theorem, a hundred head of cattle were slaughtered in celebration.

In the first part of the *Zhōubì suànjīng*, a particular case of the theorem is given: 'Gōu width three, Gǔ height four, hypotenuse length five', i.e. $a = 3$, $b = 4$, then $c = 5$. This is a particular example of the theorem of Pythagoras described above. It also says in the book 'Emperor Yǔ (禹) can rule the country because of the existence of this Gōugǔ theorem'. This indicates that the Gōugǔ theorem and in particular its special cases were probably mastered and applied by people very early on. In the commentaries of Zhào Shuǎng (趙爽) on the Zhōubì suànjīng it says 'Emperor Yǔ quells floods, he deepens rivers and streams, observes the shape of mountains and valleys, surveys the high and low places, relieves the greatest calamities and saves the people from

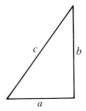

Fig. 2.2 Gōugǔ shape.

danger. He leads the floods east into the sea and ensures no flooding or drowning. This is made possible because of the Gōugǔ theorem . . .'. From this report one can infer that when the legendary Emperor Yǔ put down the floods, the Gōugǔ theorem was used.

From the latter part of the *Zhōubǐ suànjīng*, where the theory of the covering heaven (Gài Tiān theory) is discussed, one can find applications of the general Gōugǔ theorem (and no longer just special cases). In it there is an example: 'Gōu, Gǔ each multiplied by itself and added and then taking the square root, we get the hypotenuse.' That is to say,

$$c = \sqrt{(a^2 + b^2)}$$

The Gài Tiān (covering heaven) theory in the *Zhōubǐ suànjīng* uses the Gōugǔ theorem to calculate various items. For instance, it is there recorded that at the summer solstice an eight [Chinese] foot high stake casts a shadow six feet long. According to the hypothesis in the *Zhōubǐ suànjīng* we have: If the stake is moved from north to south, for every thousand miles the shadow changes by one [Chinese] inch. (One Chinese foot = 10 Chinese inches.) Now the shadow is six feet and after going south for 60 000 miles the sun is directly overhead, which means the stake has no shadow. According to the proportion of the shadow being six and the length of the stake eight, the sun should be eighty thousand miles high. It says in the *Zhōubǐ suànjīng*: 'The distance of the hypotenuse from the sun to the gauge treating the length directly under the sun as a shadow (Gōu) 60 000 miles long and the height of the sun, 80 000 miles as the Gǔ, then by the Gōugǔ [theorem], each multiplied by itself and added and the square root taken, one obtains the distance from the sun to the gauge as 100 000 miles'; that is $\sqrt{[60\ 000)^2 + (80\ 000)^2]}$ = 100 000 (see Fig. 2.3).

From a mathematical point of view one can see that the above calculation is quite accurate. However, in the empirical situation the result does not accord with the facts. First of all the assumption 'For every thousand miles the

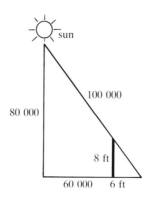

Fig. 2.3 Shadow gauge measurement.

shadow changes by one inch' is wrong. Lǐ Chūnfēng (李淳風) of the Táng
Dynasty (618–907 AD) pointed out this error. Second, the surface of the
earth is not a plane but is the surface of a sphere, while the calculation above
was made assuming that the earth is flat. In the *Zhōubì suànjīng* there are
many more calculations connected with the surveying of heavenly bodies and
deductions from them using the Gōugǔ theorem but, for the reasons stated
above, from a practical point of view they are wrong.

However, the use of the Gōugǔ theorem and two similar right-angled
triangles to survey heights, depths and distances on the surface of the earth is
quite accurate. This sort of surveying, using the knowledge of the Gōugǔ
theorem, is recorded in the *Zhōubì suànjīng*.

In the first part of the *Zhōubì suànjīng* there is a passage that runs: 'The
Duke of Zhōu asked: "... May I ask how to use the gnomon?" The Grand
Prefect (Shāng Gāo, 商高) replied: "Align the gnomon with the plumb line
to determine the horizontal, lay down the gnomon to find the height, reverse
the gnomon to find the depth, lay the gnomon flat to determine the
distance"'. That is to say (see Fig. 2.4(a)), on the line CD, D is weighted so as
to make CD a vertical line. Now align the gnomon with the vertical line so as
to see that AB is horizontal. Then one can use the gnomon to survey heights
and reverse it to find depths. As in Fig. 2.4(b, c), if the lengths of AD, CD,
and AB are known then since corresponding sides of similar triangles are
proportional, one can solve for the height or depth of BE. Similarly, using the
same principle as for surveying heights and depths, one can use the gnomon
laid flat to survey the distance between two points. Although the *Zhōubì*

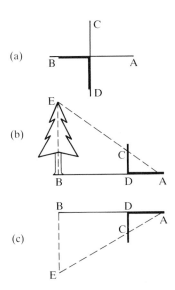

Fig. 2.4 Using the gnomon.

suànjīng only appeared as a book in about the first century BC, it is quite possible that the methods of surveying heights, depths and distances using the Gōugǔ theorem date from much earlier.

4 Calculations with fractions in the *Zhōubì suànjīng*

Besides the Gōugǔ theorem and Pythagorean surveying, the *Zhōubì suànjīng* also contains some relatively complicated calculations with fractions. Certain parts of the contents of the *Zhōubì suànjīng* also indicate that computations with fractions had reached a very high level.

All these complicated manipulations of fractions were absolutely necessary for computing the calendar and for astronomy. The Gài Tiān (covering heaven) theory in the *Zhōubì suànjīng* has the same method as the *Quarter Remainder Calendar* (四分曆, *Sìfēn lì*) for calculating the calendar in ancient times: they reckoned that one year had a length of $365\frac{1}{4}$ days (the sun changes its position by $1°$ each day as observed from earth) and they also reckoned that there should be 7 extra lunar months in 19 years, so every year there should be, on average, $12\frac{7}{19}$ lunar months and therefore the number of days in each lunar month should be:

$$365\frac{1}{4} \div 12\frac{7}{19} = 29\frac{499}{940}.$$

In the last section of the latter part of the Zhōubì suànjīng the method of division for the above fractions is recorded. ·

Now it is known that in each month there are $29\frac{499}{940}$ days, so the average angle moved by the moon each day is $13\frac{7}{19}°$. So to find out the position of the moon after 12 [lunar] months requires even more complicated calculations using fractions. In fact this is equivalent to finding the remainder in the following calculation:

$$29\frac{499}{940} \times 12 \times 13\frac{7}{19} \div 365\frac{1}{4}.$$

Of course it is not necessary to use division: one can instead use successive subtraction. The remainder is $354\frac{6612}{17860}$. In the *Zhōubì suànjīng*, the angle traversed by the moon in a leap year (that is, in a 13 lunar month year) and in an average year of $12\frac{7}{19}$ lunar months are also calculated and all these calculations involve computations with complicated fractions.

No doubt these calculations with fractions were done with counting rods. But although there are very complicated calculations with fractions used in the *Zhōubì suànjīng*, there is no explanation, in the book, of the system used for calculating with fractions. We find a discussion of the system of calculation in the *Jiǔzhāng suànshù* (九章算術 , *Nine Chapters on the Mathematical Art*), some time after the appearance of the *Zhōubì suànjīng*. Details of this are given in the next section.

2.2 The Nine Chapters on the Mathematical Art (*Jiǔzhāng suànshù*, 九章算術)

1 A brief description of the contents

Although the *Zhōubì suànjīng* includes quite advanced mathematics, its main purpose was to present the knowledge acquired in the study of astronomy, so it is not a work specifically on mathematics. The earliest specialized mathematics writing in China that survives to the present day is the *Jiǔzhāng suànshù* (九章算術, *Nine Chapters on the Mathematical Art*). A German translation has been made by Vogel (1968). A Russian translation has been made by Berezkina (1957) and Wang Ling is making an English translation (see Needham 1959 Vol. 3, p. 24).

The gradual development of ancient Chinese mathematics in the Zhōu and Qín periods and the further developments in the Hàn dynasty accumulated until they formed a complete system. The *Nine Chapters on the Mathematical Art* shows that this system had been established. It constitutes a consummation and, at the same time, a work representative of the development of ancient Chinese mathematics from the Zhōu and Qín to the Hàn dynasties (*c.* 11th century BC–220 AD). The *Nine Chapters on the Mathematical Art* had an extremely important influence on the later development of Chinese mathematics and indeed formed the basis for that development.

The *Nine Chapters on the Mathematical Art* is compiled in question and answer form. It contains a total of 246 problems and is divided into nine chapters. After giving one or more problems the book solves the problems using some particular method. One can see from this style of approach that, essentially, the inductive method is being used. On the one hand the problems can be used as worked examples after the reader has understood the topic and on the other one can use the method of solution to solve practical problems.

This problem-and-solution form had an immense influence on later mathematical works in China. Indeed the mathematical writings in ancient China adopted this form of presentation throughout.

The *Nine Chapters on the Mathematical Art* has nine chapters and each chapter has a specific title. Each chapter gives the method of calculation for one or more types of particular example.

Chapter I is called 'Field measurement' (方田, Fāng tián). The central theme is the calculation of the areas of cultivated land. The character 方 (fāng) is the unit for measuring areas; it means 'square unit', 田 (tián) means 'field' and 方田 (fāng tián) means 'to calculate how many square units a field contains'. In addition this chapter contains a detailed discussion of computations with fractions.

Chapter II is called 'Cereals' (粟米, Sù mǐ, literally millet and rice). It discusses various problems to do with proportions and in particular it is concerned with proportions for the exchange of cereals.

Chapter III is called 'Distribution by proportion' (衰分, Cuī fèn). 衰 Cuī means 'by proportion', 分 (fèn) means 'to distribute'. What is discussed are problems on proportional distribution.

Chapter IV is called 'What width?' (少廣, Shǎo guǎng). 少 (shǎo) means 'how much' and 廣 (guǎng) means 'width'. Shǎo guǎng means, given the area or volume, to find the length of a side. In this chapter the methods for finding square and cube roots are also explained.

Chapter V is 'Construction consultations' (商功, Shāng gōng). 商 (shāng) means to discuss or negotiate, 功 (gōng) means construction. This chapter deals with various kinds of calculations for constructions. It is mainly about the calculation of the volumes of various shapes of solid.

Chapter VI, 'Fair taxes' (均輸, Jūn shū) is about calculating how to distribute grain and corvée labour (see p. 272) in the best way according to the size of the population and the distances between places.

Chapter VII, 'Excess and deficiency' (盈不足, Yíng bù zú), is about the use of the method of false position for solving some difficult problems. Let us take the first problem in chapter VII as an example. The original text says: 'Consider a group of people purchasing. Each person contributes 8, and 3 are left over; 7 are contributed, 4 is the deficit.' It asks: 'How many people and what is the price?' There are two hypotheses. If p is the price and n is the number of people, the first hypothesis gives $8n = p + 3$, the second gives $7n = p - 4$. This is a problem typical of the 'Excess and deficiency' chapter.

Chapter VIII, 'Rectangular Arrays' (方程, Fāng chéng), is about equations. It discusses problems on simultaneous linear equations and it also discusses the concepts of positive and negative numbers and the methods of addition and subtraction of positive and negative numbers.

Chapter IX is on 'Gōugǔ (勾股). It discusses the Gōugǔ theorem and problems on similar right-angled triangles. In this chapter general methods of solving quadratic equations are also introduced.

The contents of the *Nine Chapters on the Mathematical Art* are comprehensive and interesting and at the same time they are closely connected with practical life. The topics that are closely connected with real life reflect the collective wisdom and abilities of the people of ancient China. But who is the author of this outstanding work, and when was this material assembled into a book? Even at the present time we are unable to give a precise and definitive answer, but according to the material presently available we can conclude that this work dates, at the latest, from about the first century AD (in the middle of the Eastern Hàn Dynasty).

There is a passage in the preface by Liú Huī (劉徽) (263 AD) to the *Nine Chapters on the Mathematical Art* that runs as follows:

It is said that the Duke of Zhōu formulated the nine rites and that consequently there was the calculation of the nines. The popularizing of the calculation of the nines led to the *Nine Chapters*. The tyrant Qín Emperor burned all the books [in 213 BC] and the classics were either lost or destroyed. It was not until later that

Marquis Zhāng Cāng (張蒼) of the Hàn and the Minister of Agriculture (Gěng Shòuchāng, 耿壽昌) became famous for their ability to calculate. Because the old text had a lot of lacunae or was incomplete and names were either missing or had been interpolated later, Marquis Zhāng and others re-edited the book and determined what was authentic or not and re-wrote part of it.

This is to say, when Liú Huī wrote his commentary on the *Nine Chapters on the Mathematical Art* in the third century AD, it was already unclear when the *Nine Chapters on the Mathematical Art* was first written and who was the first to assemble it into a book. However, from this passage one can see that the *Nine Chapters on the Mathematical Art* had passed the developmental stage of ancient mathematics in China in the Zhōu and Qín Dynasties, that knowledge had accumulated and then after the re-editing and augmentation of the text by Marquis Zhāng (?–152 BC) and Gěng Shòuchāng (Minister of Agriculture under Emperor Xuān (宣帝) of the Hàn, 73–49 BC) it was indeed a book.

So one can say that the *Nine Chapters on the Mathematical Art* was formed by the collective effort and wisdom of mathematicians of several centuries and that it was only after re-editing and augmenting by a number of people that it received its final form. As a matter of fact one can infer from certain problems and methods of calculation in the *Nine Chapters on the Mathematical Art* that the time when they first appeared was very early, while other problems and methods of calculation came into existence at a later stage.

For example, in the chapter 'Field measurement', 240 square paces is taken as equal to one acre. This was the measure used in the Warring States period and the Qín Dynasty (475–201 BC). The Hàn Dynasty, which replaced the Qín Dynasty, retained this measurement. In the chapter 'Distribution by proportion' there are the names of five ranks of the nobility in the Hàn Dynasty, such as 'Duke' (公士 , Gōngshì) and 'Marquis' (上造 , Shàngzào). This nomenclature came into existence in the Qín Dynasty. The chapter *Fair Taxes* (均輸 , Jūn shū) may have been written after the well-known historical event when Emperor Wǔ (武帝) of the Hàn Dynasty accepted the suggestion of his minister Sāng Hóngyáng (桑弘羊) to establish the official post of Jūn shū (the Fair Taxes official) in 104 BC (see also Section 2.2.3 below).

The *Nine Chapters on the Mathematical Art* is not recorded in the *Book of Arts and Crafts* in the *History of the Hàn Dynasty*. Now the *Book of Arts and Crafts*, which is based on the work of Liú Xīn (劉歆) was written towards the end of the first century BC by Bān Gù (班固). This means that about the first century BC there was probably no such book as the *Nine Chapters on the Mathematical Art*. But around 50 AD the famous scholar Zhèng Zhòng (鄭眾), in interpreting the term 'Calculations of nines' (九數 , Jiǔ shù) contained in the chapter on the official positions of Dì Guān (地官) and Bǎo Shì (保氏) in the *Zhōu Rites* (周禮 , *Zhōu Lǐ*), recorded the names of 'Field measurement', 'Cereals', 'Distribution by proportion', 'What width?', 'Construction consultations', 'Fair taxes', 'Excess and deficiency,' etc.

Therefore the contents and title of the *Nine Chapters on the Mathematical Art* must all have been completed by then.

So we can say that at the latest by the first century AD the *Nine Chapters on the Mathematical Art* had been written in the form we now have. When Liú Huī (劉徽) of Wèi in the Three Kingdoms period (220–280 AD), Lǐ Chūngfēng (李淳風) of the Táng Dynasty (618–907 AD), and others collated and commented on the book, they did not make a lot of changes in it.

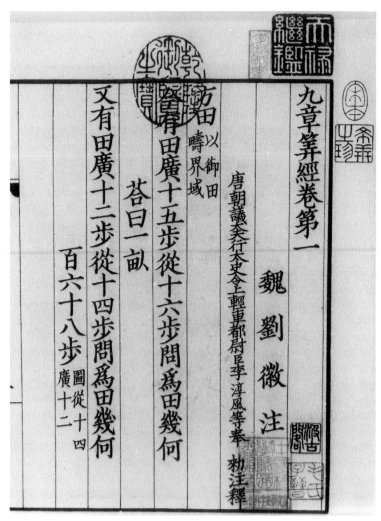

Fig. 2.5 The opening of Chapter 1 of the *Nine Chapters on the Mathematical Art* (Southern Sòng edition, in the Shànghǎi library).

Between that time and the present lie nearly two thousand years of history.

The oldest surviving edition is part of the woodblock printing of the Southern Sòng Dynasty. Only the first five chapters survive. The date of publication is shortly after 1213 AD. It is now stored in the Shànghǎi library (see Fig. 2.5). Most other editions are based on the *Complete Library of the Four Branches of Literature* (四庫全書, *Sìkù quán shū*), edited by Dài Zhèn (戴震) of the Qīng Dynasty, who copied it from the *Great Encyclopaedia of the Yǒng-lè Reign Period* (永樂大典, *Yǒng-lè dà diǎn*) of the Míng Dynasty. It is called the Dài edition.

The *Nine Chapters on the Mathematical Art* has been used as a mathematics textbook throughout the dynasties. In the Táng and Sòng Dynasties (618–1279 AD) it was made a standard textbook by the government (see below). And in fact it was a very frequently used book. The *Nine Chapters on the Mathematical Art* is also the longest surviving and most influential Chinese mathematical work ever written. There are many mathematicians of later generations who started their own research by first writing commentaries on the *Nine Chapters on the Mathematical Art*. The most famous commentaries are those of Liú Huī (劉徽) in 263 AD and Lǐ Chūngfēng (李淳風) in 656 AD. The famous mathematician Zǔ Chōngzhī (祖冲之, 429–500 AD) of the North and South dynasties also wrote a commentary on the *Nine Chapters on the Mathematical Art*, but unfortunately it has been lost.

The *Nine Chapters on the Mathematical Art* also circulated in Japan and Korea at one time and had a great deal of influence on the development of mathematics in those countries.

Finally, this important work has been noticed by the international scientific community.

2 Achievements in Arithmetic

Below we shall present the important results in the *Nine Chapters on the Mathematical Art* in the areas of arithmetic, geometry, and algebra. These results are not only splendid achievements in the ancient history of Chinese mathematics, but are also remarkable in the context of the world history of mathematics.

The achievements in arithmetic in the *Nine Chapters on the Mathematical Art* may be summarized under four headings: the systematic treatment of arithmetic operations with fractions, various types of problems on proportions, 'Excess and deficiency' problems, and other difficult problems in mathematics.

In the previous section we mentioned that very complicated calculations with fractions had been used in the *Zhōubì suànjīng*. However, that work lacked a systematic way of discussing those results. In the chapter 'Field measurement' the first 18 problems systematically discuss simplifying fractions, common denominators, comparing two fractions with different denominators, and the addition, subtraction, multiplication, and division of fractions. The methods discussed in the book are quite similar in principle to present-day methods of calculating with fractions.

Counting rods were used for calculating with fractions. The method of representing fractions using counting rods has its origins in the method of division using counting rods. For example, $123 \div 7 = 17\frac{4}{7}$ is the final result upon division using counting rods as in the following diagram:

− π	17
ⅠⅠⅠⅠ	4
π	7

The upper row is the quotient, 17, the middle row is the remainder, 4, which is the numerator of the fractional part and the bottom row is the divisor, 7, which is the denominator of the fractional part.

If the denominator and numerator had a common factor then the method of successive subtraction was used to find the largest common factor. The present-day method of successive division in modern mathematics and known in the West as the Euclidean algorithm evolved from the method of successive subtraction. This is the process by which to find the greatest common factor of numbers a and b (where $a > b$) one divides b into a (or keeps subtracting b from a) getting a remainder $r_1 < b$. One then repeats the process with b and r_1 replacing a and b. Continuing in this way one eventually obtains $r_{n+1} = 0$. Then r_n is the greatest common divisor.

Addition and subtraction of fractions requires there to be a common denominator. Finding a common denominator requires finding a common multiple. In the chapter 'Field measurement' the product of all the denominators is usually taken as the common denominator. In the chapter 'What width?', however, there are examples using the least common multiple. Thus problem number 6 shows the calculation of the following sum:

$$1 + \frac{1}{2} + \frac{1}{3} + \frac{1}{4} + \frac{1}{5} + \frac{1}{6} + \frac{1}{7}$$

$$= \frac{420}{420} + \frac{210}{420} + \frac{140}{420} + \frac{105}{420} + \frac{84}{420} + \frac{70}{420} + \frac{60}{420} = \frac{1089}{420}.$$

The common denominator 420 is the least common multiple of all the denominators.

In the West the 13th century AD Italian Leonardo Pisano (Fibonacci) is generally recognized as the first person to mention the lowest common multiple. However when he is compared with the *Nine Chapters on the Mathematical Art* it is clear that they were much earlier.

Multiplying fractions is done by multiplying the numerators and then multiplying the denominators. This is exactly the same as the modern method.

In division, however, a common denominator is found so as to make the divisor and the dividendum have the same denominator. Then the quotient is obtained by taking the numerator of the divisor as the denominator and the numerator of the dividendum as its numerator. This method of division is exemplified by

$$\frac{b}{a} \div \frac{d}{c} = \frac{bc}{ac} \times \frac{ad}{ac} = \frac{bc}{ad}$$

The other method of dividing fractions (by inverting the divisor and then multiplying) can be found in the commentary on the *Nine Chapters on the Mathematical Art* by Liŭ Huĭ in the year 263 AD. That is to use

$$\frac{b}{a} \div \frac{d}{c} = \frac{b}{a} \times \frac{c}{d}.$$

The *Nine Chapters on the Mathematical Art* is the earliest written work in the world that systematically discusses the manipulation of fractions. A similar sort of systematic discussion did not appear in India until the seventh century AD, and even later in Europe.

In the chapters 'Cereals', 'Distribution by proportion', and 'Fair taxes' in the *Nine Chapters on the Mathematical Art* there are various types of problems on proportions. For example, at the very beginning of the chapter 'Cereals' there is a list concerned with exchanging various types of cereal: 'Millet 50, unpolished rice 30, polished rice 27 . . .'. The first problem in 'Cereals' is: 'We have millet occupying 1 dǒu (斗 = 10 l), how much polished rice?' That is to say, if the exchange rate is 50 measures of millet for 30 measures of polished rice, how much polished rice do we get for 1 dǒu of millet? In other words, solve

$$50 : 30 = 1 \text{ dǒu} : x.$$

The answer given in the *Nine Chapters on the Mathematical Art* is Required number = (given number × rate for wanted cereal) ÷ rate for cereal in hand.

The problems in the chapter 'Distribution by proportion' are similar to the modern calculations of distribution by proportion. For example, there is a problem of dividing five head of deer killed hunting between officials of five ranks in the ratio 5:4:3:2:1; this is problem number one in the chapter 'Distribution by proportion'. This method of division was frequently needed in calculating taxes and corvée labour proportionately.

The problems in the 'Fair taxes' chapter involve combinations of problems from the 'Distribution by proportion' chapter and problems on proportions, thus giving compound proportional problems. Emperor Wǔ of the Hàn Dynasty (who came to the throne in 140 BC and reigned until 87 BC) used the plan suggested by Sāng Hóngyáng (桑弘羊) and established a Fair Taxes official (均輸 , Jūn shū) whose job it was to determine taxes and corvée labour. He worked these out according to the size of the population in the provinces and principalities (that is, in direct proportion to the size of the population) and the distance from the capital (that is, in inverse proportion to the distance). The methods of the 'Fair taxes' chapter are used in solving problems of this type.

The method of calculation suggested in the chapter 'Excess and Deficiency' is also a creation of ancient Chinese mathematics. To find the number of people and the price in the problem we mentioned above (p. 34): 'Each person contributes 8, 3 are left over; 7 are contributed, 4 deficit' is a typical problem in the 'Excess and Deficiency' chapter. With two hypotheses there are three possible cases to consider assuming there is either an excess or a deficit:

(a) one in excess, the other in deficit;
(b) one in excess or deficit, the other accurate; and
(c) both in excess or both in deficit.

In the chapter 'Excess and deficiency' the formulae for dealing with these three situations are given separately. If each person contributes a_1 implies b_1 is the excess (or deficit) and if each person contributes a_2 implies b_2 is the excess (or deficit) then the price can be calculated using the formula

$$x = \frac{a_2 b_1 - a_1 b_2}{a_1 - a_2}.$$

It is also possible to solve relatively more complicated problems using the methods of 'Excess and deficiency'. What is required is to make two hypotheses while knowing the results for these hypotheses and then one can deduce the required answer from the above formula. This two-hypothesis method was widely used in the Middle Ages in Europe, where it is commonly known as the Method of Double False Position. The first European to describe this method was the Italian mathematician Fibonacci in the 13th century AD. He called this method and rule of *elchataym*. This is derived from the Arabic name al-khaṭā'ain, which literally means 'double false', and sounds similar to the Arabic name used for China between the eleventh and thirteenth centuries Khiṭāi (Cathay). Nevertheless it is highly likely that this method of 'Excess and deficiency' passed into the West through the Arab world. (For further comments on this point see Needham 1959 Vol. 3, p. 118.)

3 Achievements in geometry

The problems on the calculation of areas in the *Nine Chapters on the Mathematical Art* are mainly collected in Chapter I, 'Field measurement'. We list the methods of calculation below by means of formulae in modern notation. (With the diagrams below we include the original names, 'square field', 'triangular field', etc. in brackets.)

1. Square (square field)

$$S = a^2.$$

(S is the area in this and the other formulae below.)

2. Rectangle (wide field, straight field)

$$S = ab.$$

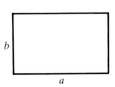

3. Triangle (triangular field)

$$S = \tfrac{1}{2}ab.$$

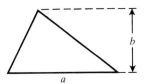

4. Trapezium (slanting field, dustpan-shaped field)

$$S = \tfrac{1}{2}(a + b)h.$$

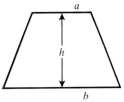

5. Circle (circular field)

$$S = \frac{P}{2} \cdot \frac{D}{2}.$$

(P = perimeter, D = diameter of base circle).

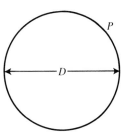

6. Segment of a circle (arc field)

$$S \doteqdot \tfrac{1}{2}(CV + V^2)$$

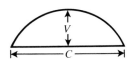

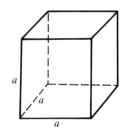

7. Segment of a sphere (domed garden field)

$$S \doteqdot \tfrac{1}{4}PD.$$

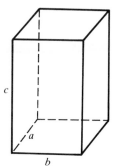

8. Annulus (ring field)

$$S = \tfrac{1}{2}(Q + P)d$$

where Q = inner circumference, P = outer circumference, and d = difference of radii.

In the list above, items 6 and 7 give approximate values. In the calculations involving circles the ratio of circumference to diameter uses 'circumference 3, diameter 1', so π is taken as 3.

Most of the calculations of volumes of familiar bodies are collected in the chapter 'Construction Consultations'. The following types are listed:

(a) Cube (square garrison tower)

$$V = a^3$$

(V is the volume here and below).

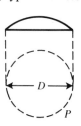

(b) Rectangular parallelepiped (Square storehouse)

$$V \doteqdot abc.$$

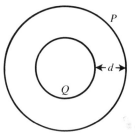

(c) Frustum of prism (Fort, earthwork, bank, ditch, moat, drain)

$$V = \tfrac{1}{2}(a + b)hl$$

(a = upper width, b = lower width, h = height, l = length).

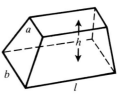

(d) Pyramid (square cone, roof, 陽馬 , yáng mǎ)

$$V = \tfrac{1}{3}a^2h$$

(a = side of base square, h = height).

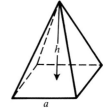

(e) Frustum (square pavilion)

$$V = \tfrac{1}{3}(a^2 + b^2 + ab)h$$

(a = side of upper square, b = side of lower square, h = height).

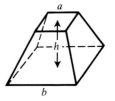

(f) Prism (diagonally cut rectangular parallelepiped)

$$V = \tfrac{1}{2}abh$$

(a = one side of base, b = other side of base, h = height).

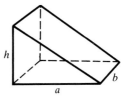

(g) Tetrahedron (turtle's shoulder joint, 鱉臑 , biē 'nào, diagonally cut prism)

$$V = \tfrac{1}{6}abh.$$

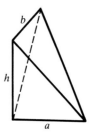

(h) Prism (tomb entrance tunnel sloping down into the ground)

$$V = \tfrac{1}{6}(a + b + c)hl.$$

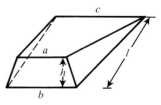

(i) Rectangular platform (haystack, pool, burial chamber)

$$V = \tfrac{1}{6}[(2a + c)b + (2c + a)d]h.$$

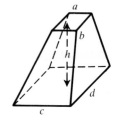

(j) Prism (roof thatch)

$$V = \tfrac{1}{6}(2b + a)ch$$

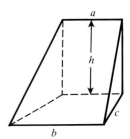

(k) Cylinder (circular fort)

$$V = \tfrac{1}{12}P^2h$$

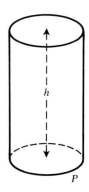

(P = circumference of circle).

If π is the ratio of the circumference of a circle to its diameter and we set $\pi = 3$, then the above formula from the *Nine Chapters on the Mathematical Art* is equivalent to the present-day formula

$$V = \pi r^2h$$

where *r* is the radius.

(l) Circular platform (Circular pavilion)

$$V = \tfrac{1}{36}(LP + L^2 + P^2)h$$

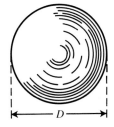

(*L* = circumference of upper circle, *P* = circumference of lower circle).

Taking $\pi = 3$ and r_1, r_2 as the radii of the upper and lower circles the above formula is equivalent to

$$V = \frac{\pi h}{3}(r_1 r_2 + r_1^2 + r_2^2)$$

(m) Circular cone (millet container, rice container, soy container)

$$V = \tfrac{1}{36}P^2 h$$

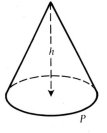

(*P* = circumference of base circle).

In the above example, again taking $\pi = 3$ and taking *r* as the radius of the base circle, the formula is equivalent to

$$V = \frac{h}{3}\pi r^2.$$

(n) Sphere (solid circle, ball)

$$V = \tfrac{9}{16}D^3$$

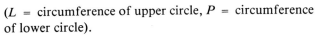

(*D* = diameter of sphere).

The 'solid circle calculation' can be found in the chapter 'What Width?'. The formula for finding the volume of a sphere is not accurate. In the chapter 'What Width?' it is claimed that the ratio of the areas of a circle and its circumscribed square is 3:4. Taking $\pi = 3$ this 3:4 ratio is correct. However, using the ratio of the volumes of a sphere and its enveloping cylinder as 3:4 is not accurate, so the volume of the sphere, which was calculated as $\tfrac{9}{16}$ of the enveloping cube, is incorrect.

It is clear that the problems of calculating areas and volumes in the *Nine*

Chapters on the Mathematical Art are practical examples from those times in surveying the sizes of fields and calculating the amount of earth in the construction of fortifications, embankments, moats and ditches, and in the construction of various types of storehouses, etc. The methods of calculation were closely connected with the day-to-day needs of those times.

4   Achievements in algebra

The main achievements in the field of algebra in the *Nine Chapters on the Mathematical Art* are: the method for the solution of a system of linear equations; the introduction of the concepts of positive and negative numbers and the methods for adding and subtracting positive and negative numbers; the methods for extracting square roots and cube roots; and the method of solution for a certain class of quadratic equations. We shall deal with these in turn below.

There are a total of 18 problems in Chapter VIII, 'Rectangular arrays', all concerned with systems of linear equations. Among these are eight problems involving two unknowns, six problems with three unknowns, two problems with four unknowns, one problem with five unknowns and one problem involving an indeterminate system of equations (five equations with six unknowns). The method used in the *Nine Chapters on the Mathematical Art* for the solution of these problems is called the 'method of rectangular arrays'. This method of calculation is known as 'rectangular arrays' because of the way the counting rods are arranged. 方 (fāng) indicates that the shape of the array is rectangular, while 程 (chéng) means a step-by-step calculation. Thus 方程 (fāng chéng) was used in ancient times as the term for the method of solving simultaneous linear equations. (Nowadays in China fāng chéng means something slightly different, namely equations.)

This method of solution may be explained as follows. As an example, in problem 1 in the chapter 'Rectangular arrays' the original statement is: 'Top-grade ears of rice three bundles, medium grade ears of rice two bundles, low-grade ears of rice one bundle, makes 39 dǒu (斗) [of rice by volume]; top-grade ears of rice two bundles, medium-grade ears of rice three bundles, low-grade ears of rice one bundle, makes 34 dǒu; top-grade ears of rice one bundle, medium-grade ears of rice two bundles, low-grade ears of rice three bundles, makes 26 dǒu. How many dǒu are there in a bundle of top-grade, medium-grade, low-grade ears of rice?' In present-day notation this is equivalent to solving:

$$3x + 2y + z = 39 \tag{2.1}$$
$$2x + 3y + z = 34 \tag{2.2}$$
$$x + 2y + 3z = 26 \tag{2.3}$$

The first step is to present the coefficients of each unknown using counting rods in columns from right to left as follows:

Top-grade ears of rice	I	II	III
Medium-grade ears of rice	II	III	II
Low-grade ears of rice	III	I	I
Total produce	=T	≡IIII	≡IIIII
	(3)	(2)	(1)

For the convenience of the modern reader we have changed the arrangement of the counting rods from right, middle, left (three column-equations) into row form as top, middle, and bottom (three row-equations). We have also adopted the usual Hindu–Arabic numerals in place of counting rods. The counting rod arrangement above is then equivalent to the following matrix of coefficients:

$$\begin{pmatrix} 3 & 2 & 1 & 39 \\ 2 & 3 & 1 & 34 \\ 1 & 2 & 3 & 36 \end{pmatrix}$$

Now use the coefficient, 3, of the first unknown in the first row to multiply the second row. Subtract from the resulting values twice the corresponding entries in the first row. This results in eliminating the coefficient of the first entry in the second row. So we have:

$$\begin{pmatrix} 3 & 2 & 1 & 39 \\ 6 & 9 & 3 & 102 \\ 1 & 2 & 3 & 26 \end{pmatrix}$$ (Three times second row.)

$$\begin{pmatrix} 3 & 2 & 1 & 39 \\ 0 & 5 & 1 & 24 \\ 1 & 2 & 3 & 26 \end{pmatrix}$$ (From the resulting row 2 subtract twice the corresponding entries in row 1.)

Similarly use the coefficient, 3, of the first unknown in row 1 to multiply row 3; subtract the corresponding entries in row 1. This makes the coefficient of the first unknown in row 3 zero. So we obtain:

$$\begin{pmatrix} 3 & 2 & 1 & 39 \\ 0 & 5 & 1 & 24 \\ 0 & 4 & 8 & 39 \end{pmatrix}$$

Next use the coefficient, 5, of the middle unknown in the second row to multiply the third row and then subtract the corresponding entries in the second row four times in succession (i.e. subtract four times the second row from the third) and thus obtain:

$$\begin{pmatrix} 3 & 2 & 1 & 39 \\ 0 & 5 & 1 & 24 \\ 0 & 0 & 36 & 99 \end{pmatrix}$$

The last row is equivalent to the equation

$$36z = 99,$$

so we get $z = 2\frac{3}{4}$.

The middle row: 0, 5, 1, 24 in the above matrix is then equivalent to

$$5y + z = 24.$$

Substitute $z = 2\frac{3}{4}$ in this equation, obtaining

$$5y = 24 - 2\frac{3}{4} = 24 - \frac{11}{4}$$

$$= \frac{24 \times 4 - 11}{4},$$

therefore,

$$y = \frac{(24 \times 4 - 11) \div 5}{4} = 4\frac{1}{4}.$$

Finally by substitution we get $x = 9\frac{1}{4}$.

One can see that this method of eliminating unknowns is essentially the same as the usual method used in present-day algebra. Yet 2000 years ago, in the *Nine Chapters on the Mathematical Art*, this method was already known as a systematic method of solving a system of linear equations. This is one of the most outstanding inventions of mathematics in ancient China. This was done at least 1500 years or so before it was done in the West. It is generally recognized that the earliest equivalent in Europe of the method of solving linear simultaneous equations in China was developed by the French mathematician Buteo in the sixteenth century.

The concepts of positive and negative numbers and the method of calculating their sums and differences are also introduced in the chapter 'Rectangular arrays'. In certain problems the amount sold is treated as positive (because of receiving money), the amount spent in purchasing as negative (because of paying out). Similarly a money balance is positive, a deficit negative. Where calculations with cereals are concerned what is added is positive, what is taken out is negative. The pair of characters 正 (zhēng) positive and 負 (fù) negative were first used colloquially at that time and are still used in this way so they have a history of about 2000 years. Astronomy in those days also employed the notions of positive and negative. In the astronomical observations and calculations used in ancient China the characters 強 (qiáng) strong and 弱 (ruò) weak were often used to indicate a certain number is approximating a certain value from above or below. For example 5.1 is called '5 strong' and 4.9 is called '5 weak'. The concepts of strong–weak and positive-negative are similar. Therefore it is written in the *Calendrical Science based on the Celestial Appearances* (乾象曆, Qián xiàng lì, 178–187 AD) by Liú Hóng (劉洪):

'Strong' (強, qiáng) is plus, positive; 'weak' (弱, ruò) is minus, negative. In addition, two same signs add up while two opposite signs take away; in subtraction, two same signs take away while two opposite signs add up.

The rules of addition and subtraction for 'strong' and 'weak' discussed by Liú Hóng (劉洪) are the same as the rules for the addition and subtraction of positive and negative numbers.

The method of adding and subtracting positive and negative numbers called 'the method of positive and negative' (正負術) is introduced in the chapter 'Rectangular arrays'. The original text says:

'The method of positive and negative states: for subtracting — same signs take away, different signs add together, positive from nothing makes negative, negative from nothing makes positive; for addition — different signs take away, same signs add together, positive and nothing is positive, negative and nothing makes negative.' (Text of Problem 3 in the chapter 'Rectangular Arrays'.)

In the first half it describes the rule for subtraction, the second half is on the method of addition and the phrases 'same signs', 'different signs' point out whether to add or subtract two numbers with the same sign or different signs. 'Add together' and 'take away' indicate that the absolute values of the two numbers are to be added or subtracted. We may explain the meaning of the passage in modern terms thus: subtracting two numbers with the same sign is equivalent to subtracting their absolute values and subtracting two numbers of different signs is equivalent to adding their absolute values; taking a positive number from zero gives a negative number, taking a negative number from zero gives a positive number; to add two numbers of different signs is equivalent to subtracting their absolute values, adding two numbers of the same sign is equivalent to adding their absolute values, adding a positive number to zero gives a positive, adding a negative number to zero gives a negative. Suppose $A > B > 0$, then 'the method of positive and negative' may be represented in modern symbols as follows:

subtraction:
$$\pm A - (\pm B) = \pm (A - B),$$
$$\pm A - (\mp B) = \pm (A + B),$$
$$0 - (\pm A) = \mp A.$$

addition:
$$\pm A + (\pm B) = \pm (A + B),$$
$$\pm A + (\mp B) = \pm (A - B),$$
$$0 + (\pm A) = \pm A.$$

It is clear that the method of addition and subtraction of positive and negative numbers introduced in the chapter 'Rectangular Arrays' is entirely correct. The method for the multiplication and division of positive and negative numbers appeared considerably later, probably not until the thirteenth century AD. The rules for multiplication and division of positive and negative numbers are clearly and accurately recorded in the *Introduction to Mathematical Studies* (算學啟蒙, *Suànxué qiméng*) by Zhū Shìjié (朱世傑), of the Yuán (元) Dynasty (1303 AD).

In calculations using counting rods, people used square section or black counting rods to stand for negative numbers and triangular section or red counting rods to stand for positive numbers; some may have used diagonal arrangements of counting rods to stand for negative and straight arrangements to stand for positive numbers.

The introduction of positive and negative numbers into mathematics in ancient China is a most outstanding invention. In India the concepts of positive and negative first appeared in the seventh century in the work of Brahmegupta (620 AD, see Colebrooke 1817/1973 Chapter XVIII, p. 339) and in Europe they had to wait until the 16th or 17th century before they had an accurate knowledge of positive and negative numbers e.g. Bombelli 1572/1966, pp. 8–9 of 1966 edn.

The fundamental principle involved in extracting square and cube roots, using counting rods, in the *Nine Chapters on the Mathematical Art* is just the same as that in the method commonly used today. They all depend on the formulae:

$$(a + b)^2 = a^2 + 2ab + b^2 = a^2 + (2a + b)b$$
$$(a + b)^3 = a^3 + 3a^2b + 3ab^2 + b^3$$
$$= a^3 + [3a^2 + 3(a + b)b]b.$$

The procedure for extracting square roots using counting rods is quite similar to the method presently used. Take, for example, problem number 12 in the chapter 'What Width?'. Let us explain how to find $\sqrt{55225}$ using the procedure presented there for extracting the square root. Since, in calculating with counting rods, the layout of the counting board is constantly changing in the process of calculation, for the present we split the steps into eight diagrams and we also use Arabic numerals to represent the counting rods for the convenience of our readers.

			(a)		(b)	
Result	商	shāng				2
Given number	實	shí	5 5 2 2 5		1 5 2 2 5	
Square element	方法	fāng fǎ			2	
Carrying rod (借算)		jiè suàn	1		1	

(c)		(d)	
	2		2 3
	1 5 2 2 5		1 5 2 2 5
	4		4
	1		1

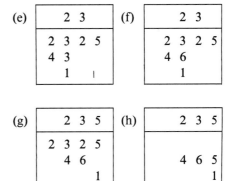

Using the symbolism of modern mathematics to explain the above calculation procedure we take the number to be N, the hundreds position of the square root as a, the tens as b and the units position as c. The formula for taking the square root of N is $\sqrt{N} = a + b + c$ and the step-by-step calculation is explained by the following diagrams:

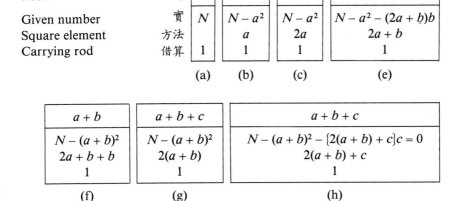

In the above diagrams we have:

(a) Lay out the number, N, whose square root is required, which is called 'Shi' (實) and in the very lowest row place one counting rod and call it the 'carrying rod' (借算 , jiè suàn).

(b) Estimate the root in the tens position then move the carrying rod to the hundreds position of the given number; and if estimating the root to be in the hundreds position move the carrying rod to the ten thousands position.

Estimate the first place in the hundreds position to be 2 (that is, a) and use 2 to multiply the carrying rod to get 2. Place this under the given number and above the carrying rod and call this the 'square element' (方法, fāng fǎ). Use 2 (that is, a) to multiply the 'square element' (also a) and then subtract from the given number to get

$$N - a.a = N - a^2 \tag{2.4}$$

(c) and (d) Continuing in order, to get the second place of the square root, double the square element, giving $2a$ (this is the first doubling of the square element) and then move the product one space backwards and move the carrying rod two spaces backwards.

One gets the second position of the number to be 3 (that is, b) in the tens position.

(e) Use 3 (that is, b) to multiply the carrying rod and then add this in to the first doubling of the square element giving $2a + b$, then use 3 (that is, b) to multiply the square element (this time it is $2a + b$) and then take away this number from the given number, giving

$$N - a^2 - (2a + b)b = N - (a + b)^2 \tag{2.5}$$

(f) Use 3 (that is, b) again to multiply the carrying rod, add it to the square element (this time it is $2a + b$), giving $(2a + b) + b = 2(a + b)$: this is the second doubling of the square element.

(g) Move the square element one space backwards after the addition and again move the carrying rod two spaces backwards, that is, move it under the units position of the given number.

Estimate the third place number in the units position to be 5 (that is, c).

(h) Use 5 (that is, c) to multiply the carrying rod and add this into the second doubling of the 'square element' giving $2(a + b) + c$, then again use 5 (that is, c) to multiply the square element [which now is $\{2(a + b) + c\}c$] and then subtract this from the given number, giving

$$N - (a + b)^2 - \{2(a + b) + c\}c = N - (a + b + c)^2 = 0. \tag{2.6}$$

The remainder is zero. From eqns (2.4), (2.5), and (2.6) the final answer is

$$\sqrt{55225} = 235.$$

The lowest row is worth noticing: the 1 which is called the 'carrying rod'. On the one hand it can be used to fix the position and on the other it can be looked on as the coefficient of x^2; and any arrangement of the counting rods reading from the lowest row to the uppermost may equally well be viewed as the coefficients of a quadratic equation. So extracting a square root may be taken on the one hand as manipulations in arithmetic for finding square roots and on the other as a general method for solving quadratic equations. In fact, in the mathematics of ancient China solving equations in mathematics grew out of the method for extracting square roots. In this the mathematicians of ancient China achieved something quite distinctive. (The details may be found below, but see also Lam Lay Yong 1980.)

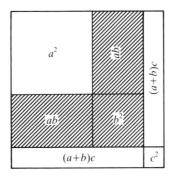

Fig. 2.6   The diagram for extracting square roots.

Problem number 20 in the Gōugǔ chapter in the *Nine Chapters on the Mathematical Art* requires the solution of the quadratic equation

$$x^2 + 34x = 71000 \qquad (x = 250).$$

In the book it simply mentions that 34 is used as the derived element (從法, Cōng fǎ) in 'the corollary to the square root method' (帶從開方法 , Dài cōng kāi fāng fǎ) and this derived method is not explained in any great detail. However, as mentioned in the previous paragraph, if we regard each configuration on the counting board as a coefficient in a quadratic equation then starting from (d) and (e) in the square root method — that is, from getting the second place number — we have the general method for solving a quadratic equation. That means that this method is a 'corollary to the square root method'. One can see this more clearly through a geometric explanation of the two methods. (See also Lam Lay Yong 1969.)

The method of extracting square roots can be given a geometric explanation as follows. If $\sqrt{N} = a + b + c$, replace this problem by the corresponding geometric idea, that is, knowing the area of the square is N, to find the length of one side. As in Fig. 2.6, after finding the first place number a, first subtract a^2 (that is, the unshaded top left hand corner) and then use $2a$ (that is, the first doubling of the square element) to get the second place b. After getting b subtract $(2a + b)b$, which is the shaded part. Finally, use $2(a + b)$ (that is, the second doubling of the square element) to find the third place c. After getting c subtract $[2(a + b) + c]c$ (which is the unshaded exterior part) from the remainder and there is nothing left.

The 'corollary to the square root method' can also be explained using the geometric method as follows. For example, suppose we want to solve the quadratic equation $x^2 + px = q$. This is equivalent to saying the area is q and we want to find one of the sides of a rectangle. From Fig. 2.7 one can see that the 'corollary to the square root method' tells us to subtract a rectangle of side p after getting one extra place of the number.

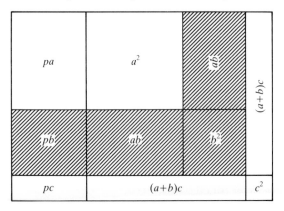

Fig. 2.7   The diagram for the corollary to the square root method.

After getting the first place number *a* we still have to subtract *pa* (the unshaded top left hand corner) after subtracting a^2.

After getting the second place number *b*, besides subtracting $(2a + b)b$ we also have to subtract *pb* (the shaded part in the middle of the left hand side). Similarly, after getting the third place *c*, besides subtracting $[2(a + b) + c]c$ we also have to subtract *pc* (the lower left unshaded part). The role of *p* in the whole procedure is quite similar to that of the 'square element' and this is why this method is called the 'derived element method'.

A square root problem involving the 'derived element method' is known as a 'corollary to the square root method'.

In comparing the geometric explanations of the above diagram let us reconsider

$$x^2 = 55255$$

(*x* = 235; this is the same as $\sqrt{55225}$)
and

$$x^2 + 34x = 63215 \qquad (x = 235).$$

The procedures for solving the above show more of the relationships between taking square roots and the 'corollary to the square root method'.

The procedure for taking the square root in the equation $x^2 = 55225$ (using present day notation, or at least the notation that used to be taught in Western high schools) is:

```
5, 5 2, 2 5 | 2   3     5
4           |_____
            | 2   43    465
  1 5 2     |
            | 2   3     5
  1 2 9     |_____
  _____ | 4   129   2325
    2 3 2 5 |
    2 3 2 5 |
  _____
        0
```

and the procedure for the 'corollary to the square root method' for the equation $x^2 + 34x = 63215$ may be presented as follows:

6 3 2 1 5	2	3	5
4 6 8	2	4 3	4 6 5
1 6 4 1 5	+ 3 4	+ 3 4	+ 3 4
1 3 9 2	2 3 4	4 6 4	4 9 9
2 4 9 5	× 2	× 3	× 5
2 4 9 5	4 6 8	1 3 9 2	2 4 9 5
0			

From the diagram one can see here that $a = 200$, $b = 30$, $c = 5$, $p = 34$. From the workings of the 'corollary to the square root method' one can also see that after getting to each place number one always adds the 'derived element' (34) then multiplies and subtracts.

The method of extracting cube roots in the *Nine Chapters on the Mathematical Art* also uses counting rods for the calculation procedure. Again the steps involve various changes of configuration on the counting board until one arrives at the result. So for finding $\sqrt[3]{N} = a + b + c$ one can explain the counting rod procedure in terms of the following figures (a)–(f). In them the top row is the partly extracted cube root, the second row is Shi (the given number), the following rows are the 'upper element' (上法, shàng fǎ), the 'lower element' (下法, xià fǎ) and the 'carrying rod':

Result	商		a	a	$a + b$
Given number	實	N	$N - a^3$	$N - a^3$	$N - a^3 - \{3a^2 + (3a + b)b\}b$
Upper element	上法		a^2	$3a^2$	$3a^2 + (3a + b)b$
Lower element	下法		a	$3a$	$3a + b$
Carrying rod	(借算)	1	1	1	1
		(a)	(b)	(c)	(d)

Result	$a + b$	$a + b + c$
Given number	$N - (a + b)^3$	$N - (a + b)^3 - \{3(a + b)^2 + 3(a + b)c + c^2\}c = 0$
Upper element	$3(a + b)^2$	$3(a + b)^2 + \{3(a + b) + c\}c$
Lower element	$3(a + b)$	$3(a + b) + c$
Carrying rod	1	1
	(e)	(f)

The method of extracting cube roots can also be explained step by step in a geometric way, but a three-dimensional figure is rather complicated and we are also limited by the size of the book, so we shall omit it here.

In conclusion we may say that in ancient times the extraction of square roots worked according to the following formulae for squares:

$$N = (a + b + c + d + \ldots)^2$$
$$= a^2$$
$$+ (2a + b)b$$
$$+ [2(a + b) + c]c$$
$$+ [2(a + b + c) + d]d$$
$$+ \ldots$$

In calculations using these formulae we reverse the procedure when extracting the root. This method of calculation is the same as the modern one. Taking the cube root is based on the following formulae for cubes:

$$N = (a + b + c + d + \ldots)^3$$
$$= a^3$$
$$+ (3a^2 + 3ab + b^2)b$$
$$+ \{[(3a^2 + 3ab + b^2) + (3ab + 2b^2)]$$
$$+ [3(a + b)c + c^2]\}c$$
$$+ \{[3(a + b)^2 + 3(a + b)c + c^2]$$
$$+ [3(a + b)c + 2c^2] + [3(a + b + c)d + d^2]\}d$$
$$+ \ldots$$

Again each step in the calculation is reversed in finding the cube root.

2.3   A mathematical work on bamboo strips found at Zhāngjiāshān

In Section 2.2 we discussed what has long been regarded as the earliest specialized mathematical writing from ancient China: the *Nine Chapters on the Mathematical Art*. However the situation has changed recently because in 1984 a new mathematical work was excavated.

In the latter part of 1983 and the beginning of 1984 a lot of bamboo strips were unearthed from three tombs (M. 247, M. 249 and M. 258) of the Western Hàn Dynasty at Zhāngjiāshān (張家山) near Jiānglíng (江陵) in Húběi (湖北) province. According to the brief report of the excavations these writings consist of the *Law of the Hàn Dynasty*, books on the *Art of War*, *On the Pulse, Chinese Traditional Gymnastics, Calendars*, etc. Most important of all for the history of mathematics in ancient China, a mathematical work was excavated at the same time from tomb M. 247. The name of this mathematical work, *A Book on Arithmetic* (算數書, *Suàn shù shū*) was written on the back of one of the bamboo strips comprising the book.

According to the archaeologists, tomb M. 247 at Zhāngjiāshān was built in the time of Empress Lǚ (呂后, 187–180 BC) or early in the reign of Emperor Wén (文帝, 179–157 BC). Clearly the mathematical writing in *A Book on Arithmetic* dates from the first half of the second century BC or earlier.

Since they had been buried under the earth for a long time a number of the bamboo strips from the Zhāngjiāshān Hàn tombs were broken or damaged. The work of repairing and restoring the order is very difficult.

From the brief report of the excavation we learn that the mathematical writing in *A Book on Arithmetic* is compiled in question and answer form just as in the *Nine Chapters*. More than sixty types of calculation are included. Some parts give the methods of calculation, for example, 'Increase and decrease' (the value of a fraction) (增減, Zēng jiǎn), 'Multiplication' (相乘, Xiāng chēng), 'Addition of Fractions' (合分, Hé fēn), etc., and some of them show the practices of economic life in the Western Hàn, for example, 'Taxes on fields' (稅田, Shuì tián), 'Field measuring' (里田, Lǐ tián — like the chapter Fāng tián in the *Nine Chapters*), 'Price of gold' (金價, Jīn jià), 'Rice calculation' (程禾, Chēng hé), and so on.

Figure 2.8 shows one of the mathematical bamboo strips from *A Book on Arithmetic*. It is an example of 'Increase and decrease' (the value of a fraction). The original text says: 'Increase and decrease. To increase the value of a fraction (one has to) increase the numerator, to decrease the value of a fraction (one has to) increase the denominator.'

From the headings in *A Book on Arithmetic* there is clearly a lot of similarity between the mathematical bamboo strips and the *Nine Chapters on the Mathematical Art*. The following example shows the original text from the two works and they are very much alike. Under the title 'What Width?' (少廣, Shǎo guǎng) there is a mathematical problem.

The problem is that there is a rectangular field whose area is 1 mǔ (畝 = 240 bù 步). The length of one side is one and a half bù (1 bù = 6 Chinese feet in the time of the Western Hàn). What is the width? The original text says:

What width? (少廣, shǎo guǎng). When one side is one and a half bù [the method of calculation is to] let one become 2 and a half become 1, add them together to obtain 3, take 3 as the divisor. Then take [the number for the area] 240 bù, let 1 become 2 again, divide by the divisor [3] and the size of the width is obtained. It is 160 bù . . .

A similar question is in the first problem of the chapter 'What width?' of the *Nine Chapters on the Mathematical Art*. The orignal text says:

There is a [rectangular] field [whose area is] 1 mǔ. When one side is one and a half bù, what is the width? The answer is 160 bù. The method of calculation is that there is a half in the later part [of the problem]. Half is $\frac{1}{2}$. Let one become 2, half become 1, add them together, obtain 3, take 3 as divisor. Then take the [area of the] field, 240 bù, make 1 become 2 again i.e. multiply by 2, and take this as the dividendum. Divide it by the divisor and the width is obtained.

Another point we have to mention here is that in the *Nine Chapters on the Mathematical Art* there is a chapter named 'Fair Taxes' (均輸, Jūn shū) and we wrote that the 'Fair taxes' chapter had been thought to have been written after the official post of Jūn Shū was established by the Emperor Wǔ (104

To increase or decrease a fraction: to increase a fraction increase its numerator; to decrease a fraction increase its denominator.

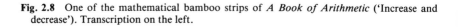

Fig. 2.8 One of the mathematical bamboo strips of *A Book of Arithmetic* ('Increase and decrease'). Transcription on the left.

BC). Now among the bamboo strips there were found the *Law of the Hàn Dynasty* (excavated from tomb M.247) and also the *Law of Fair Taxes* (均輸律 , *Jūn shū lü*). Thus the solution of mathematical problems of 'Fair taxes' must date from earlier times (before 100 BC).

The discovery of the mathematical bamboo strip *A Book on Arithmetic* indicates that there were a number of mathematical writings that had been written before the *Nine Chapters on the Mathematical Art*. In fact the *Nine Chapters* were compiled on the basis of those mathematical writings and that is why the *Nine Chapters* has become a representative classic of the mathematics of ancient China.

Further detailed research into the mathematical bamboo strips from Zhāngjiāshān is still going on.

3

The development of mathematics in China during the Wèi, Jìn, and North and South Dynasties

(221–589 AD)

3.1 The Illustrated Commentary on the Right Triangle, Circle, and Square (勾股圓方圖注 , Gōugǔ yuán fāng tú zhù) by Zhào Shuǎng (趙爽)

The appearance of the *Nine Chapters on the Mathematical Art* signalled the first integration of mathematics into a system in ancient China. New developments were built on this foundation in ancient China during the period of the Wèi, Jìn, and North and South Dynasties (third to sixth centuries AD). From the material available we can say that this development began with Zhào Shuǎng's *Commentary on the Zhōubì suànjīng* (周髀算經注 , *Zhōubì suànjīng zhù*), was continued by the subsequent *Commentary on the Nine Chapters on the Mathematical Art* (九章算術注 , *Jiǔzhāng suànshù zhù*) by Liú Huī (劉徽) and the mathematics of this period culminated in the work of the Zǔ's, father Zǔ Chōngzhī (祖冲之) and his son Zǔ Gěng (祖暅). This marked the highpoint in the course of the development of mathematics in China after the two Hàn dynasties.

Below we start by first introducing the work of Zhào Shuǎng (趙爽). Zhào Shuǎng, also called Jūn Qīng (君卿), lived about the time of the Wèi and Jìn periods (third and fourth centuries AD). His mathematical work remained within the confines of the *Commentary on the Zhōubì suànjīng* and its most valuable part is the 'Illustrated commentary on the right triangle, circle, and square' (勾股圓方圖注).

In the Ming edition of the *Zhōubì suànjīng* there is mention of 'Zhào Shuǎng of Hàn' so people decided that Zhào Shuǎng was of the Hàn people. However the *Commentary on the Zhōubì suànjīng* uses the work of Zhāng Héng (78–139 AD) *The Spiritual Constitution (or Mysterious Organization) of the Universe* (靈憲 , *Ling xiàn*) and Liú Hóng (劉洪) *Calendrical Science based on the Celestial Appearances* (乾象曆 , *Qián xiàng lì*) and, in particular, the latter work was used during the Three Kingdoms period (221–280 AD) by people in the kingdom of Wú, so one can determine that Zhào Shuǎng lived in the third or fourth century AD. The first person to say that Zhào Shuǎng was from the time of the North and South Dynasties was the Southern Sòng scholar Bào Huǎnzhī (鮑澣之).

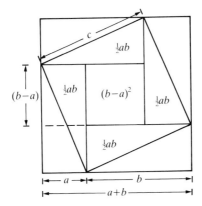

Fig. 3.1 The diagram on the hypotenuse.

The 'Illustrated commentary on the right triangle, circle, and square' can be found in the present edition of the *Zhōubì suànjīng*. The whole text is only about 500 Chinese characters. It lists 21 theorems on the four main systems involving right-angled triangles and the relations between the three sides. Using present-day notation, set the sides of the triangle adjacent to the right angle with the horizontal side, gōu, = a and the vertical side, gǔ, = b, and let the hypotenuse = c. Then the first system mentioned is equivalent to

$$\begin{cases} a^2 + b^2 = c^2, \\ c = \sqrt{(a^2 + b^2)}. \end{cases}$$

This is the same as the Gōugǔ theorem and as such already occurs in the *Zhōubì suànjīng*. The remaining three systems concern theorems on the 'diagram on the hypotenuse' (弦圖), the 'gōu gnomon' (勾實之矩) and the 'gǔ gnomon' (股實之矩).

All the diagrams from the 'Illustrated commentary on the right triangle circle and square' have been lost. One can infer however that the 'diagram on the hypotenuse' was something like Fig. 3.1. That is, the outer square has side $a + b$, the intermediate square has side c, the length of the hypotenuse, as its side, and the small square inside has $b - a$ as its side.

From Fig. 3.1 we know:

$$2ab + (b - a)^2 = c^2 \tag{3.1}$$

$$\text{and} \quad \frac{c^2 - (b - a)^2}{2} = ab = A,$$

$$b - a = B,$$

then solving

$$x^2 + Bx = A$$

one obtains

$$x = a, \text{ and } a + B = b. \tag{3.2}$$

From the diagram we know: $2c^2 - (b-a)^2 = (b+a)^2$,

therefore $$\sqrt{[2c^2 - (b-a)^2]} = b + a. \tag{3.3}$$

From the diagram we again have: $2c^2 - (a+b)^2 = (b-a)^2.$ (3.4)

From (3.3) and (3.4) we obtain

$$\frac{(b+a) + (b-a)}{2} =$$

$$\frac{\sqrt{[2c^2 - (b-a)^2]} + \sqrt{[2c^2 - (b+a)^2]}}{2} = b, \tag{3.5}$$

$$\frac{(b+a) - (b-a)}{2}$$

$$\frac{\sqrt{[2c^2 - (b-a)^2]} - \sqrt{[2c^2 - (b+a)^2]}}{2} = a. \tag{3.6}$$

There remain what are known as various systems of formulae about the 'gōu gnomon' (勾實之矩, gōu shí zhī jǔ) (see Fig. 3.2). What is called the 'gōu gnomon' is obtained as follows: take a square with each side equal to the hypotenuse of the original triangle and take away from it a smaller square (b^2, bottom left of Fig. 2.10) whose side is equal to the length of the gōu and then the remaining part of the figure has area $c^2 - b^2 = a^2$. This is the square of the gōu. The shape is then called the *gōu-shí* (勾實); its shape is like the *jǔ* (矩), the carpenter's square which can be found in Chapter 1 of this book. That is why this is called the *jǔ of the gōu-shí* (勾實之矩, the gōu-gnomon).

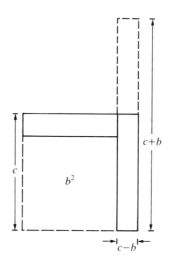

Fig. 3.2 The gōu-gnomon.

Its area accordingly is $a^2 = c^2 - b^2$ and one can immediately deduce that:

$$\text{area of gōu-gnomon} = a^2 = c^2 - b^2 = (c + b)(c - b).$$

From this we have

$$\sqrt{[c^2 - (c + b)(c - b)]} = \sqrt{[c^2 - a^2]} = b. \tag{3.7}$$

Again $$(c + b) - (c - b) = 2b.$$

Let $$c - b = x,$$

then $$x(2b + x) = (c - b)(c + b) = a^2,$$

therefore $x^2 + 2bx = a^2$, and this equation has solution

$$x = c - b. \tag{3.8}$$

Now from $(c + b)(c - b) = a^2$ we know that

$$\left.\begin{array}{l} \dfrac{a^2}{c + b} = c - b \\[2mm] \dfrac{a^2}{c - b} = c + b. \end{array}\right\} \tag{3.9}$$

Hence we have

$$\begin{aligned} c &= \frac{(c + b) + (c - b)}{2} \\[2mm] &= \frac{1}{2}\left[(c + b) + \frac{a^2}{c + b} \right] \\[2mm] &= \frac{(c + b)^2 + a^2}{2(c + b)}, \end{aligned} \tag{3.10}$$

$$b = \frac{(c + b)^2 - a^2}{2(c + b)}.$$

Now, if we let $(c + b)(c + b) = A$, $(c + b) + (c - b) = B$, then, because

$$[(c + b) + (c - b)]^2 - 4(c + b)(c - b)$$
$$= [(c + b) - (c - b)]^2$$

it follows that $$(c + b) - (c - b) = \sqrt{(B^2 - 4A)}$$

and hence:

$$c - b = \frac{1}{2}[B - \sqrt{(B^2 - 4A)}]. \tag{3.11}$$

Using formula (3.11) it is possible to solve the problem of finding two numbers whose sum and product are given. Let two numbers (a, b) have sum B and product A, then the problem is equivalent to solving $x(B - x) = A$ which is $-x^2 + Bx = A$, where the coefficient of the leading term is negative

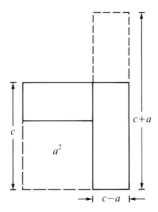

Fig. 3.3   The gŭ-gnomon.

(so this is different from the method derived from that for extracting square roots).

The gnomon called the 'gŭ-gnomon' (股實之矩 , gŭ shí zhī jŭ) is obtained as follows: take a square with each side equal to the hypotenuse of the original triangle and take away from it a smaller square (a^2, bottom left of Fig. 3.3), whose side is equal to the length of the gŭ. The remaining part of the figure then has area ($c - a$). Therefore

$$\text{area of gŭ-gnomon} = (c + a)(c - a),$$

hence we have

$$\sqrt{[c^2 - (c + a)(c - a)]} = \sqrt{(c^2 - b^2)} = a. \tag{3.12}$$

Let $c - a = x,$

then solving $\left. \begin{aligned} x^2 + 2\,ax &= b^2, \\ x &= c - a. \end{aligned} \right\} \tag{3.13}$

one gets

Again, from $(c + a)(c - a) = b^2$, we know

$$\left. \begin{aligned} \frac{b^2}{c + a} &= c - a \\ \frac{b^2}{c - a} &= c + a. \end{aligned} \right\} \tag{3.14}$$

Consequently, we have

$$\left. \begin{aligned} c &= \frac{1}{2}[(c + a) + (c - a)] = \frac{(c + a)^2 + b^2}{2(c + a)} \\ a &= \frac{1}{2}[(c + a) - (c - a)] = \frac{(c + a)^2 - b^2}{2(c + a)}. \end{aligned} \right\} \tag{3.15}$$

Amongst all these equations, eq (3.11) is worthy of note; it is equivalent to solving $-x^2 + Bx = A$. This is a quadratic equation with a leading term with a negative coefficient. The method of solution is similar to the method used in India and the Arab world in the Middle Ages. It is a method which derived from 'arranging squares' (配平方 , pèi píng fāng) in geometric patterns. This way of using direct representation by geometric figures is a new kind of method. It was also used extensively in the *Commentary on the Nine Chapters on the Mathematical Art* by Liú Huī (details below).

3.2 The contribution of Liú Huī (劉徽)

1   The method of circle division

We now introduce another mathematician besides Zhào Shuǎng from the Three Kingdoms period (220–280 AD): we consider the work of Liú Huī (劉徽).

Almost no material concerning the life and background of Liú Huī survives to the present day, but from the *Commentary on the Nine Chapters on the Mathematical Art* and the *Sea Island Mathematical Manual* (海島算經 , *Hǎidǎo suànjīng*), which have survived, we have no hesitation in saying that Liú Huī was a great mathematician. The two above-mentioned books record a great deal of his important and creative work.

Liú Huī did not mention the date in his *Commentary on the Nine Chapters on the Mathematical Art* but in the *Memoir on the calendar in the History of the Sui Dynasty* (隋書 · 律曆志 , *Sui shū lù lì zhì*) there is a sentence which says that in the fourth year of the Jìng Yuán (景元) reign of the King Chén Liu (陳留王) of Wèi (that is, 263 AD) Liú Huī wrote a commentary on the *Nine Chapters on the Mathematical Art*. Again, from some of the passages in the *Commentary on the Nine Chapters on the Mathematical Art* we can say that he lived in the third century AD in the Wèi dynasty. And because work from the 'Illustrated commentary on the right triangle, circle and square' by Zhào Shuǎng is used in the *Commentary on the Nine Chapters on the Mathematical Art* (for Problems 5 and 11 in the Gōugǔ chapter of the *Nine Chapters*) it is possible to say that he lived later than Zhào Shuǎng.

After the addition of Liú Huī's explanations and commentary there were no further changes made to the *Nine Chapters on the Mathematical Art* and the work has survived in this form to the present. With the commentary by Liú Huī the *Nine Chapters* became more comprehensive. We know that the *Nine Chapters* listed general methods of calculation but the explanations and discussions are very brief. The commentary by Liú Huī manages to make up for those deficiencies. We can go one step further and say that the commentary and explanations for the *Nine Chapters* gave brief proofs for the various types of calculation and verified the accuracy of the calculations.

The *Commentary on the Nine Chapters on the Mathematical Art* by Liú Huī touches on many aspects and includes a great deal of inventive work. The 'method of circle division' is one of the most important of its achievements. The 'method of circle division' produced a new method of calculating π, the ratio of the circumference to the diameter. Mathematicians in ancient China made outstanding contributions to the calculation of π and Liú Huī's 'method of circle division' has an important place among these.

In the very beginning ancient mathematicians in China took the ratio of circumference to diameter as 'Circumference 3, diameter 1', that is to say, π = 3. [There never has been a Chinese word for π, it is always referred to as 'the circle circumference ratio' (圓周率 , yuán zhōu lü).] Although a lot of mathematicians and astronomers before Liú Huī had used different values for π, nobody had suggested a systematic and scientific method for calculating π. For example, when Liú Xīn (劉歆) made a bronze hú (斛, a vessel of about 100 litres capacity), for Wáng Mǎng (王莽, c.10 AD) he used $\pi \doteqdot$ 3.1547; Zhāng Héng (張衡, 78–139 AD) used $\pi = \sqrt{10} \doteqdot 3.16$ to find the volume of a sphere; Wáng Fán (王蕃, 219–257 AD) used $\pi = 142/45 =$ 3.1556. In his 'method of circle division', Liú Huī was the first to use inscribed regular polygons and then, by increasing the number of sides, gradually approach the circle and thence find π. The 'method of circle division' is given in the commentary after problem 32 in the 'Field measurement' chapter of the *Nine Chapters*.

Problem 32 in the 'Field Measurement' chapter concerns the calculation of the area of a circle. In the original text of the *Nine Chapters on the Mathematical Art* it gives the rule for calculating the area of a circle as 'half the circumference multiplied by half the diameter'. Let S be the area of the circle and r the radius, then the formula in the *Nine Chapters* is equivalent to

$$S = r \cdot \frac{2 \pi r}{2} \left(= \pi r^2 \right).$$

This formula is exact but because the value π = 3 was used the area computed was not accurate.

We can summarize Liú Huī's 'method of circle division' as follows:

1. First Liú Huī points out that using the value π = 3 does not give the area of the circle but of the regular dodecagon (12-sided polygon) inscribed in the circle. This area is less than the correct answer.

2. Liú Huī started his calculations with an inscribed regular hexagon and then increased the number of sides, doubling the number of sides each time, and calculating the areas of the regular dodecagon, then the regular 24-gon, then the regular 48-gon, then the regular 96-gon and so on, thus making the area calculated approach the area of the circle.

3. Liú Huī pointed out that knowing the side of the regular hexagon one can find the area of the regular dodecagon and that knowing the side of the regular dodecagon it is possible to find the area of the regular 24-gon and so on. Thus in general, knowing the side of the regular $2n$-gon it is possible to find the area of the regular $4n$-gon.

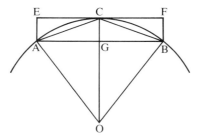

Fig. 3.4 The method of circle division (1).

As in Fig. 3.4 let AB be the side of the regular $2n$-gon and let its length be l_{2n}. Then following the rule in the *Nine Chapters*, half the perimeter multiplied by half the diameter gives the area of the regular $4n$-gon, so the area

$$S_{4n} = r \cdot \frac{2nl_{2n}}{2} = 2n \cdot \frac{r \cdot l_{2n}}{2}$$

From Fig. 3.4 we know that $\dfrac{r \cdot l_{2n}}{2}$ is the area of the diamond ACBO and this diamond shape is one of $2n$ identical parts of the regular $4n$-gon. So by the above explanation, Liú Huī's conclusion is correct.

4. Liú Huī pointed out that the area (S) of the circle is bounded by

$$S_{2n} < S < S_{2n} + (S_{2n} - S_n) \, (n = 6, 12, 24, \ldots).$$

From Fig. 3.4, $(S_{2n} - S_n) = n(\triangle ACG + \triangle BCG)$, which clearly establishes the above inequalities.

5. From a hexagon one can calculate one side of a regular dodecagon and from a regular dodecagon find one side of a regular 24-gon . . . and in general from a $2n$-gon one can find the length of one of the sides of a regular $2n$-sided polygon. Liú Huī had repeatedly applied the Gōugǔ theorem. The general formula can be written as:

$$l_{4n} = \sqrt{\left\{ r - \sqrt{\left[r^2 - \left(\frac{l_{2n}}{2} \right)^2 \right]} \right\}^2 + \left(\frac{l_{2n}}{2} \right)^2}$$

From Fig. 3.4, the Gōugǔ formula gives $\sqrt{\left[r^2 - \left(\dfrac{l_{2n}}{2} \right)^2 \right]} = $ OG, and $r - $ OG $= $ CG. For the right angled triangle ACG, knowing CG and AG $\left(= \dfrac{l_{2n}}{2} \right)$, the above formula then follows by using the Gōugǔ theorem again.

6. Starting from the hexagon inscribed in a circle of radius 1 Chinese foot, Liú Huī calculated the length of the side of a regular 96-gon and then the area of the regular 192-gon S_{192}, finding

$$S_{192} = 3.14 \frac{64}{625} \text{ square feet.}$$

This is equivalent to finding $\pi = 3.141024$.

For practical calculations Liú Huï used $\pi = 3.14$ or $\pi = \dfrac{157}{50}$.

7. Liú Huï recognized that the area S_{192} is not the final answer and indicated that it was still possible to continue to 'divide'. In the original text of the *Commentary on the Nine Chapters on the Mathematical Art* his words are recorded as follows: 'The finer it cuts, the smaller the loss; cut after cut until no more cuts, then it coincides with the circle. Thus there is no loss.' That is to say: the greater the number of sides the closer the area of the inscribed regular polygon approaches the area of the circle. When the number of sides increases to infinity then the limit of the area of the regular polygon is the area of the circle (cf. Wagner 1979, p. 173.)

Calculating the ratio of the circumference of a circle to its diameter by using the areas of regular polygons with an increasing number of sides was known about the third century BC. It was first done by the ancient Greek mathematician Archimedes. However Archimedes used both an inscribed and a circumscribed polygon for his calculation whereas Liú Huï used only an inscribed polygon, so his method was simpler. It should also be noted that Liú Huï's method was not influenced by Archimedes. He obtained it independently.

The calculation of the ratio of π improved to an accuracy of six decimal places from the time of Liú Huï's work up to the time of Zǔ Chōngzhï (429–500 AD). (Details in the next section.) The calculation of π is one of the main mathematical achievements of ancient China.

2   Other achievements in the *Commentary on the Nine Chapters on the Mathematical Art*

The concept of limit

We can see from the discussion in the previous section that in the process of calculating π Liú Huï used some notion of limit. Liú Huï indicated similar ideas in other parts of his *Commentary on the Nine Chapters on the Mathematical Art*.

Liú Huï also thought that the calculation of the area of a segment of a circle should be accomplished by means of a method like that of circle division. He thought that the original method given in the *Nine Chapters* only applied in the calculation of the area of the segment which is exactly a semicircle, and even there the calculation was not for the segment of the circle but rather for the inscribed regular dodecagon. Using the method of circle division would give a much better approximation.

First of all, in Fig. 3.5 v_0 is the length of the 'arrow' or sagitta and c_0 is the

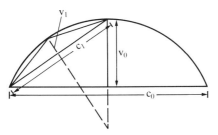

Fig. 3.5 The method of circle division (2).

length of the 'bowstring' or chord. These can be used to find the radius of the circle (in the 'Gōugŭ chapter of the *Nine Chapters on the Mathematical Art* there are similar problems) and then by the method of circle division the 'short arrows' and 'short bowstrings' of the $\frac{1}{2}, \frac{1}{4}, \frac{1}{8}, \ldots$ parts of the segment can be found. Letting this sequence of 'short arrows' and 'short bowstrings' be $c_1, c_2, \ldots, v_1, v_2, \ldots$, then clearly

$$\text{area of the segment} = \frac{1}{2}(c_0v_0 + 2c_1v_1 + 4c_2v_2 + \ldots).$$

In the *Commentary on the Nine Chapters* Liú Huī is recorded as saying: 'Cuts and further cuts, getting extremely fine, then string times arrow gives a very close answer.' This is another use of the concept of limit. The commentary goes on to record further remarks of Liú Huī: 'Although the calculation is very complicated as far as the arithmetic is concerned, we should get the final answer. However, when measuring fields take a rough estimate which is a good approximation.' That is to say that in calculating the areas of actual fields only an approximate value is required. Liú Huī fully realized the import of the theory and the practice of these approximate calculations.

The commentary after problem 16 in the 'What width? chapter in the *Nine Chapters* says: '. . . the extraction goes on and on, the decimal fraction getting smaller and smaller, . . . although there is something left over it is not worth mentioning.' And again in the commentary on problem 15 in the 'Construction consultations' chapter in the *Nine Chapters* it says 'Divided in half it is a small amount and the remainder is very tiny and when it becomes extremely tiny (微 , wēi) then it has no form (形 , xíng). From this point of view why should we bother about the remainder?' (See also Wagner 1979, p. 173.).

Where Liú Huī discussed the residue when extracting non-integral square roots and finding the area of irregular figures we should use the concept of limit. The concept of limit derives from the analysis of infinitesimals — it is the basic concept in higher mathematics. In Europe, although this concept was developed very early (for example, in the time of the ancient Greeks), it was only in the seventeenth century that there was further refinement and development. Liú Huī was the first mathematician of ancient China to use the concept of limit to solve mathematical problems.

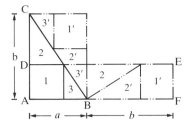

Fig. 3.6   Patchwork solution of geometrical problems (1).

The methods of dividing, combining, moving, and compensating in geometry

In Liú Huï's preface to the *Nine Chapters on the Mathematical Art* Liú Huï says that his commentary 'is to analyze by words, to explain by diagrams, to give concise but complete principles and should be easily understood but not repetitious'. Indeed, the two main aspects of Liú Huï's commentary are the use of words to explain, and of diagrams and models for discussion. In the *Commentary on the Nine Chapters* Liú Huï systematically developed the method of using diagrams and models and other geometrical methods in explaining various mathematical problems.

In the solution of problems on areas Liú Huï used the method of combining figures of various kinds. In fact, this is equivalent to the method of translation and rotation used in plane geometry today.

For example, in the problem of finding the length of the side of a square inscribed in a right-angled triangle whose sides (a, b) adjacent to the right angle are given, when commenting on problem 15 of the *Gōugǔ* chapter of the *Nine Chapters*, Liú Huï used a diagram like Fig. 3.6 to prove the original formula in the *Nine Chapters*:

$$x = \frac{ab}{a + b}.$$

This formula is correct, since $ab = 2\,\triangle ABC$, which can be rearranged to form the rectangle ADEF as in Fig. 3.6.

Again, in finding the radius (r) of the circle inscribed in a right-angled triangle whose sides (a, b) adjacent to the right angle are given, when commenting on problem 16 of the *Gōugǔ* chapter in the *Nine Chapters*, Liú Huï used Fig. 3.7 to explain the formula

$$\text{diameter } D = \frac{2ab}{a + b + c}$$

This is correct — $ab = 2\,\triangle ABC$ and as in Fig. 3.7, by rearranging $2\,\triangle ABC$ it is possible to form a rectangle with sides D and $a + b + c$. In the original

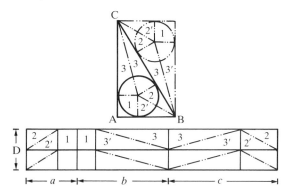

Fig. 3.7 Patchwork solution of geometrical problems (2).

text of Liú Hui's commentary it says: 'Use diagrams on small paper, cut diagonally through the intersections, rearrange them together, combine them to form particular shapes, . . .'. On the surface this method of rearranging geometric figures looks as though it is only a method of explanation and discussion. In fact it is a form of proof.

Liú Hui's calculation of the volume of solid objects is similar to his calculation of areas, as there he also uses the methods of synthetic geometry to rearrange various types of three-dimensional pieces then called 'qi (棋)' (chess pieces).

There are four types of commonly used qi. (see Fig. 3.8):

(a) *Cube* (including rectangular parallelepiped).

(b) *Prism* (half of a cube cut diagonally).

(c) *Pyramid* (cutting a prism diagonally gives a pyramid and a tetrahedron, pyramid = $\frac{1}{3}$ of a cube). This special pyramid is called a yáng-mǎ (陽馬). This is also the name given to a corner of a hip-gabled roof (see Wagner 1979, p. 175).

(d) *Tetrahedron* (volume = $\frac{1}{6}$ of a cube). This special tetrahedron is called a biē'naò (鱉臑), which literally means a turtle's shoulder joint (see Wagner 1979, p. 166).

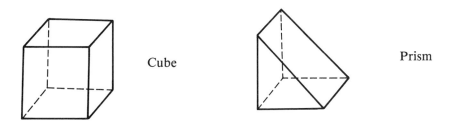

Cube

Prism

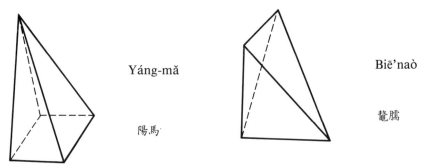

Relatively complicated volumes can be calculated by putting a number of these qì together. For example in order to find the volume of a square platform (frustum of a pyramid) when the sides of the top and bottom squares and the height are known, in his commentary on Problem 10 of the 'Construction consultations' chapter of the *Nine Chapters*, Liú Huī used 'one cube in the middle, four prisms on four sides, and four pyramids (yáng-mǎ) on four corners' to form a square platform as in Fig. 3.8. In proving that the original formula in the *Nine Chapters*

$$\text{volume } V = \frac{1}{3} h (ab + a^2 + b^2)$$

is correct, he wrote the following commentary: 'top and bottom side [a,b] multiply each other, . . .; multiply by the height [h], to get the middle cube and the four prisms on four sides. Bottom side multiplied by itself, . . ., multiplied by the height, . . . to make one middle cube, two sets of four prisms and three sets of yáng-mǎ. The upper side multiplied by itself, multiplied by the height, . . . gives another central cube.'

This passage says:

	Middle cube	Square prism	Yáng-mǎ
$a.b.h.$ is equivalent to	1	4	—
$b^2.h$ is equivalent to	1	8	12
$a^2.h$ is equivalent to	1	—	—
Total of 27 qì	3	12	12

Liú Huī's commentary also says: 'The number of qì used: three cubes, twelve each of prisms and pyramids (yáng-mǎ), a total of 27 qì. Use the three and the twelve assembled together to make three square platforms. Q E F (驗矣 , yàn yǐ).' That is to say, using 27 qì one can form three square platforms, so multiplying by $\frac{1}{3}$ one gets the volume of the square platform, thus proving the accuracy of the formula.

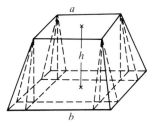

Fig. 3.8 Patchwork method for volumes.

Liú Huī uses similar methods to calculate the volume of various other bodies. For example:

Fodder loft (芻甍, chú méng): use 'two *prisms* in the middle, two yáng-mǎ at two ends' (Fig. 3.9).

Drain (羨除, xiàn chú): 'two or four *tetrahedra* with a prism in between' (patchwork diagram omitted).

Square awl (方錐, fāng zhuī): 'four *pyramids*' (patchwork diagram omitted).

Fodder boy (芻童, chú tóng): 'two *cubes*, eight *prisms*, four *pyramids*' (patchwork diagram omitted) etc.

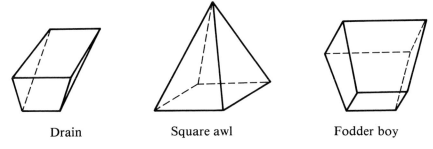

Drain   Square awl   Fodder boy

In the problems on calculating volume Liú Huī also uses plane sections. Obviously this too is a straightforward geometrical method. For example, in Problem 11 in the 'Construction consultations' chapter, calculating the volume of a circular pavilion (圓亭, yuán tái), Liú Huī used the method of comparing it with a circumscribed square pavilion (Fig. 3.10). In his commentary he says: 'To find the volume of a circular pavilion from that of a square one, we can start from the ratio of a circular area to a square one. Taking the circle as 3 multiply by the ratio to the square 4, then get the volume of the circular platform.' Thus he is using:

circular pavilion : square pavilion
= circle area : area of circumscribing square,
= $3(\pi)$: 4.

As far as the volume of a sphere is concerned, the *Nine Chapters* claims

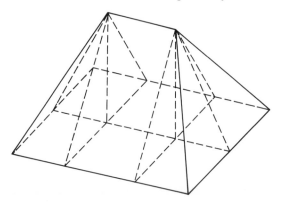

Fig. 3.9   Patchwork method for volume of a chú méng (刍甍).

that the ratio of the volume of the sphere to the volume of its circumscribing
cylinder is equal to the area of the circular cross-section to the area of the
circumscribing square, which is not correct. Liú Huī pointed out this error
and he discovered that the ratio of the volume of the sphere to the volume of
the 'two square umbrellas' (牟合方盖 , móu hé fāng gài) is π:4, where 'two
square umbrellas' are formed by the intersection of two solid cylinders that
circumscribe the sphere and whose axes intersect at right angles. However,
Liú Huī did not go on to calculate the volume of 'two square umbrellas'. He
made the following comments towards the solution of this problem:
'Wanting to get the volume of the shape, afraid of losing the truth, not daring
to guess, I wait for a capable man to solve it.' The way Liú Huī set up this
problem 'waiting for a capable man to solve it' opened the way for his
successors and his approach is worth noting.

Liú Huī started on the calculation of the volume of the sphere and it was
completed more than two hundred years later by the great mathematicians Zǔ
Chōngzhī and his son in the period of the North and South dynasties (details
in the next section).

Furthermore, Liú Huī also developed the method of decimal place value
notation and opened the way to the use of the decimal point. In the commentary by Liú Huī on Problem 6 of the 'What width?' chapter of the *Nine
Chapters* it says: 'The tiny number without a name is the numerator, retreat
one space using 10 as the denominator and retreat another using 100 as the
denominator; keep on going back, the subdivisions becoming very fine.'
That is to say the smaller fractions after fēn (分) = $\frac{1}{10}$, lǐ (厘) = $\frac{1}{100}$, háo
(毫) = $\frac{1}{1000}$, sī (絲) = $\frac{1}{10000}$, hū (忽) = $\frac{1}{100000}$, have no name but each time
one can 'go back one place using 10 as denominator, then go back again using
100 as denominator'. Thus 3.1415 can be expanded as

$$3 + \frac{1}{10} + \frac{4}{100} + \frac{1}{1000} + \frac{5}{10000}.$$

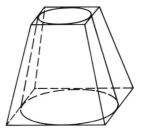

Fig. 3.10 Finding the volume of a circular pavilion.

In addition he treated the notions of positive and negative numbers. In the commentary by Liú Huī on Problem 3 in the 'Rectangular arrays' chapter of the *Nine Chapters* it says: 'If the gain and loss in two calculations are opposite then use positive and negative to name them.' But due to the limitations of space in this book it is impossible to treat these in detail.

3 The Sea Island Mathematical Manual (海島算經 , *Hǎidǎo suànjīng*)

Besides writing the commentary on the *Nine Chapters on the Mathematical Art*, Liú Huī has another work which has survived to the present: this is the *Sea Island Mathematical Manual*. It has been translated into French by L. van Hée (1920, 1932).

The contents of the *Sea Island Mathematical Manual* concern the 'method of double differences' (重差 , chóng chā), a method used in surveying. This book is not an independent work, for in the original preface to his *Commentary on the Nine Chapters on the Mathematical Art*, Liú Huī says:

. . . in researching the *Nine Calculations* (九數 , Jiǔ shù) [*Nine Chapters*] there is a term 'double differences' (重差 , chóng chā) . . . for finding heights or surveying great depths while knowing the distances, one must use 'double differences' . . . as well as doing the commentary I researched the meaning of the ancients and reformulated 'double differences' as an appendix to the *Gōugǔ* chapter.

From this we can see that the present book was originally part of Liú Huī's commentary explaining the method of 'double differences' and was appended to the 'Gōugǔ' chapter in the *Nine Chapters*.

By the seventh century AD, at the beginning of the Táng Dynasty, this part had been taken out of the *Nine Chapters* and had become an independent work. It is called the *Sea Island Mathematical Manual* because the first problem concerns surveying the height and distance of an island in the sea.

By the beginning of the Qīng (清) Dynasty (in the late seventeenth century) the *Sea Island Mathematical Manual* was no longer in circulation. The ordinary text that is available today was in the compilation of the

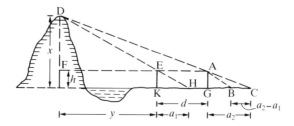

Fig. 3.11   Calculating the height and distance of a sea island.

Complete Library of the Four Branches of Literature (四庫全書 , *Sì kù quán shū*) by Dài Zhèn (戴震 , 1724–1777 AD), who copied it from the *Great Encyclopaedia of the Yǒng-Lè Reign Period*. Today's edition of the *Sea Island Mathematical Manual* contains nine problems.

The last few problems in the 'Gōugǔ' chapter of the *Nine Chapters* are problems on surveying the height of a fortified city, the height of a mountain, the depth of a well, etc. Liú Huī's 'double differences method' in the *Sea Island Mathematical Manual* is the further development of this sort of method in surveying.

Why is it called the 'double differences method'? We can explain this by summarizing Problem 1 in the *Sea Island Mathematical Manual*.

The content of Problem 1 is as follows: Observe a sea island whose height and distance are unknown. Erect two poles (表, biǎo) AG and EK, as in Fig. 3.11. The height of the poles is b feet and the distance between the two poles is d paces, while the two poles and the island are lined up in the same vertical plane. Step back a_1 paces from the front pole (EK) so as to observe the top of the pole and the top of the island in the same straight line when the eye is at ground level. Again, step back a_2 paces from the rear pole and observe the top of the pole and the top of the island in the same straight line with the eye at ground level. It is required to find the height of the island (x) and the distance between the front pole and the island (y).

The solution in the original text of the *Sea Island Mathematical Manual* is:

Use the pole height [h] multiplied by distance between poles as numerator, the differences [$a_2 - a_1$] as denominator. The result obtained added to the pole height gives the height of the island. To find the distance from the front pole to the island [y], use the distance walked forward from the front pole [a_1] multiplied by the pole distance [d] as numerator, the differences [$a_2 - a_1$] as denominator to get the number of miles from island to pole.

Using modern algebraic notation the above explanation yields the formulae:

$$x = \frac{d}{a_2 - a_1} \cdot h + h ,$$

$$y = \frac{d}{a_2 - a_1} \cdot a_1 .$$

These formulae can be proved using Fig. 3.11. Let D be the top of the island. From A construct AB//DE. Then knowing $\triangle$ABC is similar to $\triangle$ADE and also $\triangle$ACG is similar to $\triangle$ADF we have

$$\frac{AE}{BC} = \frac{d}{a_2 - a_1} = \frac{AD}{AC} = \frac{DF}{AG} = \frac{DF}{h} ,$$

and hence $x = DF + h = \dfrac{d}{a_2 - a_1} \cdot h + h$ QED.

Again because $\triangle$ EKH is similar to $\triangle$ DFE,

$$\frac{y}{a_1} = \frac{EF}{KH} = \frac{DF}{EK} = \frac{DF}{h} = \frac{\dfrac{d}{a_2 - a_1} \cdot h}{h} = \frac{d}{a_2 - a_1} ,$$

hence $y = \dfrac{d}{a_2 - a_1} \cdot a_1$ QED.

Now assume the point D is not the top of the island but the sun and that a_2 and a_1 are the lengths of the shadows of the two poles. Then using the above method it is possible to find the height of the sun from the ground. In fact, astronomers in the Western Hàn period used this method for surveying. d is the distance between the poles, which are in line with the object being surveyed, and $a_2 - a_1$ is the difference in the lengths of the shadows of the two poles. This sort of calculation requires two differences, so this method of surveying is called 'double differences'.

Since the surface of the earth is not flat but spherical, using the 'double differences method' to survey the sun gives an inaccurate value. But using the 'double differences method' to survey limited distances on the ground is highly accurate. In the *Sea Island Mathematical Manual* Liú Huī used the

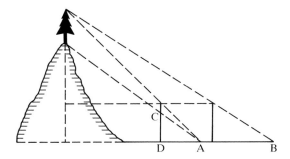

Fig. 3.12 Calculating the height of the tree on a mountain.

'double differences method' for surveying various objects on the earth. This
sort of problem requires two observations but some problems require three or
four, depending on the nature of the problem. For example, to survey the
height of a tree on a mountain requires three observations. As in Fig. 3.12 the
top of the tree must be observed from both points A and B and the base of the
tree must be observed from point A. One more datum is required than for
surveying the height and distance of the island, namely the height CD.

Again, in Problem 7 of the *Sea Island Mathematical Manual*, using the
stone at the bottom of the bank on the opposite side of a river to find the
depth of the river as in Fig. 3.13 needs two observations from point A to find
the lengths of BC and BD. It also requires two observations from point E
to get the lengths of FG and FH. So in addition to knowing the fixed distances AB and EF (where AB = EF), four observations are needed to find the
depth *x*.

Of the nine problems in the surviving text of the *Sea Island Mathematical
Manual*, three require two observations (numbers 1, 3, 4), four require three
observations (numbers 2, 5, 6, 8), and two require four observations
(numbers 7, 9). All of these calculations and observations can be illustrated
by using similar right-angled triangles and the principles of proportion.
Although this type of surveying computation did not use the concepts of trigonometrical functions, having the length of the pole and using the propor-

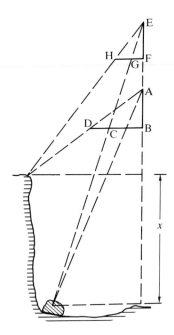

Fig. 3.13   Calculating the depth of a stone at the bottom of a river.

tions between line segments, it was just as feasible to obtain accurate results. In the preface to the *Commentary on the Nine Chapters on the Mathematical Art* Liú Huī himself says: 'For measuring heights depend on poles, for investigating depths erect gnomons; a distant object alone needs three observations, a distant object with something else needs four observations, measuring an object by sighting requires lying down, otherwise it cannot be done.' The *Sea Island Mathematical Manual* demonstrates the progress and development of mathematical surveying in ancient China.

As is well known, mathematical surveying is also the basis of map-making. In 1973 three marvels of cartography were excavated in Hàn tomb no. 3 at Mǎwángduī (馬王堆) near Chángshā (長沙), the capital of Húnán province. Judging from the archaic names of the places noted in it and the

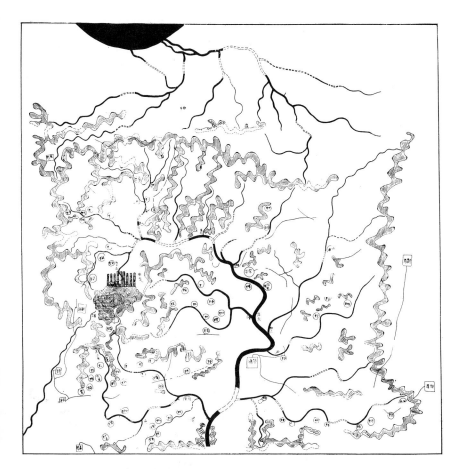

Fig. 3.14 Reconstruction of the topographical map unearthed in Han tomb no. 3 at Mǎwángduī near Chángshā. (96 cm × 96 cm) Contrary to modern practice, south is at the top.

time the tomb was built (168 BC) we are sure that these maps were made in the early years of the Western Hàn (206 BC–24 AD).

One of these three maps is a topographical map of Chángshā Guó (長沙國). Figure 3.14 is a reconstruction of it. In contrast to modern practice this map was drawn to a scale of approximately 1 : 180 000. The location of the main part of the map is fairly accurate and when the courses of waterways are compared, for example, the Shēnshǔi (深水) river (now called Xiāoshǔi, 瀟水) and its tributaries, then they largely coincide with those in a modern map. The topography of this part of Húnán province is very complicated so it is impossible to measure many distances between landmarks directly. Thus indirect measuring, the method of double differences was needed here.

The topographical map excavated at Mǎwángdui could only have been made by using the double differences method. This method of mathematical surveying, as was mentioned above, is discussed fully in Liú Huī's book, the *Sea Island Mathematical Manual*, but the method, as the Mǎwángdui map proves, was already known in the early period of the Western Hàn.

3.3 The outstanding mathematician Zǔ Chōngzhī in the period of the North and South Dynasties

1   The biographies of Zǔ Chōngzhī (祖冲之) and Zǔ Gěng (祖暅)

Zǔ Chōngzhī (429–500 AD) had Wényuǎn (文遠) as his other name. He was an outstanding mathematician in the period of the North and South Dynasties in the time of the Sòng (宋) and Qí (齊) Dynasties.

The ancestral home of Zǔ Chōngzhī was in northern China but his grandfather and father were officials of the Southern Dynasty and he must have been born in the South.

From the end of the Western Jìn Dynasty (316 AD), because there was continued war in the North, most of the population in the central plain moved south and this forced the rapid development of agriculture and the economy in the Yangtze plain. Zǔ Chōngzhī was born in this period.

For generations the Zǔ family had researched into astronomy and calendar making. Under his family's influence Zǔ Chōngzhī had a deep interest in astronomy and mathematics.

During his youth he carefully and thoroughly studied the work of Liú Xīn (劉歆), Zhāng Héng (張衡), Wáng Fán (王蕃) and Liú Huī (劉徽). He corrected a lot of their errors. Thereafter he remained immersed in research and made great contributions to many aspects of science and technology. His determination of π to an accuracy of seven decimal places is one of his most outstanding achievements.

As far as astronomy and calendar making were concerned he collected together all the available literature from antiquity to his own time as well as giving proofs and conducting experiments through his own personal observations and calculations. He pointed out a lot of serious errors in the computation of the calendar of Hé Chéngtiān (何承天) (370–447 AD) used at that time. Consequently he began constructing a new type of calendar.

In the sixth year of the Dà Míng (大明) reign of the Liú-Sòng (劉宋) Dynasty (462 AD) the new calendar was constructed. It is known as the 'Dà Míng Calendar'. At that time Zǔ Chōngzhī was 33 years old. Judging from the scientific knowledge of that period it was the best calendar available. However, the new calendar provoked a lot of objections from very important influential people in the court such as Dài Fǎxīng (戴法興). Because of fear of the influence of Dài Fǎxīng most of the court officials were afraid to give an unbiased opinion of the new calendar of Zǔ Chōngzhī. In order to defend the truth, Zǔ Chōngzhī bravely started a public debate with Dài Fǎxīng. He wrote a very famous essay entitled 'Riposte' (駁議, bó yì) in which he contested all the unreasonable claims of Dài Fǎxīng.

In fact this debate reflected the struggle between science and anti-science, between progress and conservatism, in the development of science at that time. Dài Fǎxīng and others contended that what had been passed down through history was the product of the Sages and therefore could not be amended but must remain unchanged through thousands of generations. They thought that astronomy and the making of calendars could not be altered by ordinary humans. They said: 'Zǔ's skin-deep thoughts cannot hope to penetrate'; they even went so far as to accuse Zǔ Chōngzhī of 'blaspheming against heaven and working against the Classics'. Zǔ Chōngzhī rebuked them sharply. He maintained that the motions of the sun, the moon, and the five known planets were 'not from spirits nor from ghosts' but 'have determinate form and can be worked out with numbers', and that only through careful observation and computation can one realize the ancient sages' statement that 'Summer and Winter solstices of a thousand years can easily be determined'. Zǔ Chōngzhī wrote two very famous lines in the 'Riposte'; they are 'Willing to hear and look at proofs in order to examine truth and facts' and 'Floating words and unsubstantiated abuse cannot scare me'. He hoped that both sides would present concrete evidence to determine the rights and wrongs and he was not at all afraid of unfounded rumours and abuse.

It is because of all the obstructions of this sort that the Dà Míng Calendar was not adopted until the time of the Liáng (梁) Dynasty (510 AD), ten years after Zǔ Chōngzhī's death.

Besides the science of the heavens, calendar making and mathematics, Zǔ Chōngzhī also researched into many aspects of machinery. He invented the 'south-pointing chariot' which is believed to have been a chariot with a gear mechanism that makes an arm constantly point south (see Zhou Shide 1983, p. 432) and the 'thousand-mile boat', that is to say a boat fast enough to cover a great distance in one day. He also knew the theory of music very well and edited and commented on a number of ancient texts. Besides all this, he wrote ten volumes of novels. It can be said of him that he was truly a universal scholar.

In the *Catalogue of Classical Works* in the *History of the Sui dynasty* (隋書 · 經籍志, *Sui shū · Jing ji zhi*) there are 51 books recorded of the collected works of Admiral Zŭ Chōngzhī. Unfortunately these collected works have long been lost. Of his works on mathematics the most important one must be the *Method of Interpolation* (綴術, *Zhui shù*) but there were also the *Commentary on the Methods and Essence of the Nine Chapters* (九章術義注 , *Jiǔzhāng shù yi zhù*), the *Commentary on Double Differences* (重差注 , *Chóng chā zhù*), and others. All are lost. From the material that is presently available we can only present a fraction of the work of this great mathematician of the fifth century AD. This is on aspects of the calculation of π and the volume of the sphere (see Sections 3.3.2 and 3.3.3).

Zŭ Gĕng, also known as Zŭ Gĕngzhī (祖暅之) the son of Zŭ Chōngzhī, was also an outstanding mathematician who continued the work of his father in mathematics, astronomy, and calendar making. He also further developed the achievements of his father. It was only after three memorials by Zŭ Gĕng that the 'Dà Míng calendar' of Zŭ Chōngzhī was accepted by the Liáng dynasty. In a number of catalogues from ancient times the book *Method of Interpolation* (綴術 , *Zhui shù*) is listed as the work of Zŭ Gĕng. The calculation of the volume of the sphere was preserved because of the subsequent work of Zŭ Gĕng. Zŭ Gĕng's whole life was spent in learning. According to legend he studied so hard when he was still very young that he did not even notice when it thundered; when he was thinking about problems while walking he bumped into people.

The names of the father, Zŭ Chōngzhī, and his son, are spoken of approvingly everywhere in China today, and they are also quite well known around the world.

2   On the calculation of π

According to the surviving material the pre-eminent achievement of Zŭ Chōngzhī in mathematics is his calculation of π. This well-known and outstanding achievement was recorded in the *Memoir on the calendar* in the *History of the Sui* (隋書 · 律曆志 , *Sui shū lü li zhi*)

In the original text of the *Memoir on the calendar, History of the Sui* it says: 'Ratio: circumference 3, diameter 1. This method is as full of holes as a sieve. Later Liú Xīn (劉歆), Zhāng Héng (張衡), Liú Huī (劉徽), Wáng Fán (王蕃), Pí Yánzōng (皮延宗) and others found new ratios, but these still did not give accurate bounds, . . . Zŭ Chōngzhī plugged the holes using 1 zhàng, (丈) as the diameter giving circumference upper limit as 3 zhàng, 1 foot, 4 inches, 1 fēn, 5 li, 9 háo, 2 miǎo, 7 hū; lower limit 3 zhàng, 1 foot, 4 inches, 1 fēn, 5 li, 9 háo, 2 miǎo, 6 hū; the correct number is between the upper and lower limits. Close ratio: diameter 113, circumference 355; approximate ratio: diameter 7, circumference 22.' [A zhàng (丈) is 10 Chinese feet while a chǐ (尺) is a Chinese foot of 10 Chinese inches (fēn, 分) and the succeeding measures lí (厘), háo (毫), etc. are each a tenth of the preceding one.]

From the point of view of appreciating the achievement this passage is very clear but it is obviously not sufficient to show how Zǔ Chōngzhī obtained this result or how he computed it.

Using modern notation for clarity this passage means: Circumference 3, diameter 1 ($\pi = 3$) is an inaccurate ratio. Although Liú Xīn, Zhāng Héng, Liú Huī and others set up various new ratios they were not accurate enough. Zǔ Chōngzhī calculated further to obtain a very accurate ratio. He used 1 zhàng as the diameter and divided it into 100 000 000 parts (that is, 1 zhàng = 100 000 000 parts) for the calculation and finally found that the accurate circumference should be between 3 zhàng, 1 foot, 4 inches, 1 fēn, 5 lí, 9 háo, 2 miǎo, 7 hū and 3 zhàng, 1 foot, 4 inches, 1 fēn, 5 lí, 9 háo, 2 miǎo, 6 hū. That is to say, Zǔ Chōngzhī calculated

$$3.1415926 < \pi < 3.1415927 \ .$$

This result is accurate to seven decimal places. Using this ratio for calculation, if the radius of a circle is 10 kilometres, then the error is at most a few millimetres. This is indeed very accurate.

In the world history of mathematics many mathematicians of various countries tried very hard to calculate an accurate value for π. Once a German mathematician said: 'In history the accuracy of the ratio of the circumference to the diameter calculated in a country can serve as a measure of the level of scientific development of that country at that time'. Zǔ Chōngzhī calculated the ratio of the circumference to the diameter to seven decimal places. This is a measure of the high level of development of mathematics in ancient China.

From the table below we can see the accuracy of computations of the ratio of the circumference to the diameter by various mathematicians in different countries throughout the history of the world:

Archimedes (Ancient Greece)	287?–212 BC	3.14 (accurate to 2 decimal places)
Liú Huī	263 AD	3.14 or 3.1416 (accurate to 2 or 4 decimal places)
Zǔ Chōngzhī	429–500 AD	3.1415926(7) (accurate to 7 decimal places)
al-Kashi (15th century Persia)	1427 AD	accurate to 16 decimal places
Viète (France)	1540–1603 AD	accurate to 10 decimal places

| Ludolf van Ceulen
(Germany) | 1539–1610 AD | accurate to 35 decimal
places |

Using modern computers π can be calculated to several thousand decimal places. For example, on 20 July 1959 Genuys, using an IBM704, calculated π to 16 167 decimal places (see Wrench 1960).

We can see that it was not until the fifteenth century AD that al-Kashi, the assistant of the Persian astronomer Prince Ulugh Beg, broke Zǔ Chōngzhī's record of seven decimal places' accuracy. However, this is almost 1000 years after the death of Zǔ Chōngzhī.

How Zǔ Chōngzhī calculated this result cannot be known in detail because the material given in the *Memoir on the Calendar* in the *History of Sui* is too scanty. Moreover, according to research, it is impossible to know whether there were any other new methods in existence which could have been used by Zǔ Chōngzhī apart from the method of circle-division of Liú Huǐ. In fact continuing the calculation using the method proposed by Liú Huǐ as far as the regular 24 576-gon one can obtain this result. If the inference is correct then Zǔ Chōngzhī had to perform very complicated operations on nine-digit numbers more than one hundred times (including taking square roots). One can only think this mathematician carried out a very difficult and laborious task. (However it is possible that Zǔ Chōngzhī did not do the calculation for the 6×2^{12}-sided polygon. For example in the seventeenth century the Japanese mathematician, Seki Kowa (關孝和 , 1642–1708) used the 'continuous cancellation method'. Here the number of sides is not required to increase so quickly in order to arrive at a similar value for π.)

According to the custom for calculations in those days (they were used to using fractions) Zǔ Chōngzhī suggested two fractions for the ratio of the circumference to the diameter:

a close ratio (relatively accurate) : $\pi = \dfrac{355}{113}$, which is equivalent to 3.1415929, six decimal places' accuracy and an approximate ratio :

$\pi = \dfrac{22}{7}$, which is equivalent to 3.14, two decimal places' accuracy.

355/113 is a very close fraction. This value was derived in Europe during the sixteenth century by the German mathematician Otto (1573 AD). This is more than 1000 years later than Zǔ Chōngzhī. The number $\frac{22}{7}$ is relatively simple and is convenient to use and was first found by Archimedes about 250 BC. In the history of Western mathematics it has always been taken that the value $\dfrac{355}{113}$ was first obtained by the Dutchman Anthoniszoon (1527–1607 AD) near the end of the sixteenth century AD and the ratio has been called Anthoniszoon's ratio. However a Japanese mathematical history has suggested this should be called 'Zǔ's ratio'.

3   On the calculation of the volume of the sphere

Here we introduce another important achievement of Zǔ Chōngzhī, the calculation of the volume of the sphere.

There is a passage in the essay 'Riposte' where Zǔ Chōngzhī retorted to Dài Fǎxíng: '. . . about the error on the sphere [this refers to the error in the calculation in the *Nine Chapters*] Zhāng Héng (張衡) mentioned it but left it uncorrected. . . . It is a scar on mathematics. . . . I have corrected these errors of the past, . . .'. From this we can see that before Zǔ Chōngzhī had completed the Dà Míng calendar, he had already accurately calculated the volume of the sphere. However, when Lǐ Chūngfēng (李淳風) in the Táng Dynasty used this method of computation in his commentary (on Problem 24 in the 'What Width?' chapter of the *Nine Chapters* on the method of finding the diameter of a sphere whose volume is given) he claimed it was a new method invented by Zǔ Gěng. So the calculation of the volume of the sphere may be said to be the achievement of the father and son of the Zǔ family.

In the preceding section we mentioned that in the *Nine Chapters* it was suggested that the ratio of the volume of the sphere to the volume of its circumscribing cylinder was equal to the ratio of a circle to its circumscribing square (π : 4). Liú Huī had already pointed out that this was incorrect. At the same time he also suggested the correct ratio of the volume of the sphere to the 'two square umbrellas' was equal to the ratio of the circle to the circumscribing square. However, Liú Huī was unable to calculate the volume of the 'two square umbrellas' (牟合方蓋 , móu hé fáng gài). Liú Huī had said: 'Wanting to get the volume of the shape, afraid of losing the truth, not daring to guess, I wait for a capable man'. Liú Huī raised the problem; 250 years later it was solved by the geniuses father Zǔ Chōngzhī and his son.

The method used by the father and son of the Zǔ family can be briefly described as follows. Take a small cube so that its side has length equal to the radius of the sphere which is r. This is $\frac{1}{8}$ of the cube circumscribing the sphere. As in Fig. 3.15(a), with 0 as centre and r, the radius of the original sphere as radius, construct two cylinders cutting the side and front faces of the cube. In this way the small cube is cut into four parts as in Fig. 3.15(b–e). In (b) we have exactly $\frac{1}{8}$ of what Liú Huī called 'two square umbrellas' — let us called it a 'part umbrella'. Now recombine all four parts to make a small cube and at height h take a cross-section. The bottom part is as in Fig. 3.15(f). In the right-angled triangle ABC, AB is the radius (= r), BC is the height (= h) and AC is the length (call it a) of a side of the square cross-section of the 'part umbrella'. By the Gōugǔ theorem we know

$$\overline{AC}^2 = \overline{AB}^2 - \overline{BC}^2$$

that is: $a^2 = r^2 - h^2$ = area of the cross-section of the 'part umbrella'. Let S be the sum of the areas of the gnomon consisting of the other three surfaces of the cross-section (the shaded part in Fig. 3.15(f). Then

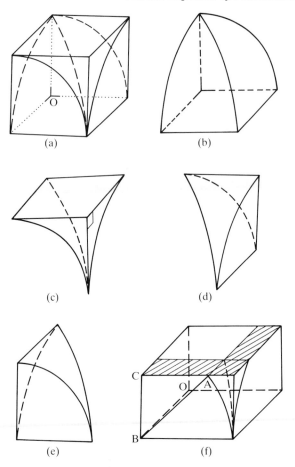

Fig. 3.15 The cross-section of the 'part umbrella'.

$$S = r^2 - a^2 = r^2 - (r^2 - h^2) = h^2.$$

Also it is very easy to see that no matter where the cross-section is taken, $(0 \leqslant h \leqslant r)$, $S = h^2$ always holds.

Now take a pyramid whose height and side $= r$, turn it upside down (i.e. point down) and compare its volume with that of the three parts *outside* the part umbrella. It is easy to determine that for all heights h $(0 \leqslant h \leqslant r)$ the area, S of the cross-section of the pyramid is equal to h^2. Using this, the father and son of the Zǔ family solved the problem. They said: 'Matching membranes, same stature, then the volume cannot be different.' 'Membrane' (冪 , mì) is to indicate the surface of the cross-section, 'stature' (勢 , shì) is to indicate the height. Thus they are saying that after cutting away the three parts from the 'part umbrella', although the shapes are not the same, still the volumes can be calculated using the surface areas of the cross-sections and the heights, and since the total surface areas of the cross-sections are equal at the same height, it follows that their volumes cannot be unequal. From this it follows that the volume of the three parts is equal to the volume of the pyramid, which is equal to $\frac{1}{3}$ the volume of the small cube. (The volume of the pyramid is in the 'Construction Consultations' chapter of the *Nine Chapters on the Mathematical Art*, see above.) Therefore the volume of the 'part umbrella' is equal to $\frac{2}{3}$ the volume of the small cube.

Next we can deduce that the volume of the whole of 'two square umbrellas' is equal to $\frac{2}{3}$ of the volume of the large cube whose side is equal to the diameter (D). Again, because the ratio of the square cross-sectional area of the 'two square umbrellas' and the inscribed sphere at the same height is just the ratio of the square to the circle, we can therefore calculate the volume (V) of the sphere:

$$V = \frac{\pi}{4} \cdot \frac{2}{3} D^3 = \frac{4}{3} \pi r^3 \, .$$

This formula we know to be correct.

In this problem the father and son of the Zǔ family applied the method of comparing the areas of cross-section to provide an exceptionally elegant solution of the problem of the volume of the sphere.

Although the problem of the volume of the sphere had been solved very early in the West by the ancient Greek mathematician, Archimedes, the work of the father Zǔ Chōngzhī and his son was done independently. In particular the method they used was particularly ingenious and remains an outstanding achievement. The theorem 'equal cross-sectional area at equal heights implies the two bodies have the same volume' is commonly known as the theorem of Cavalieri (1598–1647 AD). It is generally claimed that it was first used by the Italian mathematician Cavalieri. This is not in accordance with the facts, as Zǔ Chōngzhī used it a thousand years before Cavalieri, so it really ought to be called 'Zǔ's theorem'. (See also Lam Lay-Yong (1985).)

Using Zǔ's theorem, volumes of some very complicated bodies can be found indirectly through computing some simple volumes. In some syllabuses of solid geometry in present-day secondary schools in China Zǔ's theorem is still very important.

4

Chinese mathematics in the time of the Suí and Táng Dynasties
(581–907 AD)

4.1 Research in the method of interpolation by astronomers in the Suí and Táng Dynasties

In the several centuries from the time of the North and South Dynasties to the period of the Suí and Táng, there were noticeably significant improvements in various aspects of research in astronomy and calendar computation. In these we can, in particular, clearly distinguish complementary developments in astronomy and mathematics. The continual improvement of the calendar required far more accuracy in the methods of computation. The 'method of interpolation' or, to use more accurate terminology from modern algebra, the 'method of second differences interpolation' was first used by the astronomer Liú Zhuó (劉焯) (544–610 AD) in the Suí Dynasty.

What is the 'method of interpolation' and what is the 'method of second differences interpolation'? Let us explain. We know the intermediate values between $1, 2, 3, 4, 5, 6, \ldots$ are $1.5, 2.5, 3.5, 4.5, 5.5, \ldots$. To find them, add adjacent numbers and divide by 2, so $(2 + 3) \div 2 = 2.5$, $(3 + 4) \div 2 = 3.5$, etc. This is very simple but if we want to find the intermediate numbers between the squares $1, 4, 9, 16, 25, 36, \ldots$ of $1, 2, 3, 4, 5, 6, \ldots$ that is, from $2^2 = 4$, $3^2 = 9$, etc., it is required to find the square of 2.5, then the above method cannot be used since $(4 + 9) \div 2 = 6.5$, while $(2.5)^2$ is actually 6.25. Before taking the squares of $1, 2, 3, 4, 5, 6, \ldots$ the interval is 1 so after taking the squares $1, 4, 9, 16, 25, 36, \ldots$ the differences are called 'equal interval second order numbers'. To find the intermediate values given 'equal interval second order numbers' is not trivial but requires establishing certain formulae. Liú Zhuó was the first person to introduce the 'method of second differences interpolation' or the 'method of interpolation'. For example, given $1, 4, 9, 16, 25, 36, \ldots$ he knew how to calculate $(1.5)^2$, $(2.5)^2, \ldots$ and even $(1.7)^2$, $(2.8)^2, \ldots, (6.37)^2$. Liú Zhuó already grasped this method of interpolation. This was indeed an outstanding innovation.

The editing and making of the calendar and in particular the forecasting of solar and lunar eclipses required knowing the exact positions of the sun, the moon and the five known planets. Before the time of the Eastern Hàn Dynasty (25–220 AD), people believed that the speeds of movement of the sun, the moon, and the five known planets were equal, that is each day the angular distances moved were the same. Jiǎ Kuí (賈逵 , 92 AD), in the period of the Eastern Hàn, discovered that the movement of the moon is sometimes fast and sometimes slow. In the time of the North and South Dynasties, Zhāng Zǐxin (張子信), an astronomer of the Northern Dynasty (c. sixth century AD) observed the movement of the sun from a sea island for thirty years. Finally he discovered that the movement of the sun from the point of view of an earthbound observer is sometimes faster, sometimes slower. This is because the movement of heavenly bodies is not along a circular orbit but along an elliptical one.

Under these circumstances how does one accurately calculate the position of the sun, moon, and the planets?

It is obvious that one cannot just use observations to determine the position every hour and every minute. For example, during daylight the sun is so bright that it is impossible to see the other celestial bodies, so it is impossible to determine the relative positions of the sun and the planets. So how can we calculate the positions of the sun, moon, and the five planets between two observations? This requires the method of interpolation.

The method of interpolation first of all requires knowing much data from observations at different times. Assuming the time intervals between successive observations are equal, then this is called the 'equal time interval method of interpolation'. If the intervals between the times are not equal then it is called the 'method of interpolation with unequal time intervals'.

In what follows we use modern algebraic notation to explain these two methods of interpolation.

Suppose the constant time interval is w and at times $w, 2w, 3w, \ldots, nw,$ $\ldots$ the results of the observation are $f(w), f(2w), f(3w), \ldots, f(nw), \ldots$. If we randomly choose a time between two observations as a time between w and $2w$, we can represent it by $w + s$ (where s satisfies $0 < s < w$) and $f(w + s)$ can be found using the formula

$$f(w + s) = f(w) + s\Delta + \frac{s(s - 1)}{2!} \Delta^2$$

$$+ \frac{s(s - 1)(s - 2)}{3!} \Delta^2 + \ldots$$

The meanings of $\Delta, \Delta^2, \Delta^3, \ldots$ are given by the following. Let

$$\Delta_1' = f(2w) - f(w),$$

$$\Delta_2' = f(3w) - f(2w),$$

$$\Delta_3' = f(4w) - f(3w),$$

$$\Delta_1^2 = \Delta_2' - \Delta_1', \qquad \Delta_2^2 = \Delta_3' - \Delta_2',$$

$$\Delta_1^3 = \Delta_2^2 - \Delta_1^2$$

then the values of $\Delta, \Delta^2, \Delta^3, \ldots$ in the above formulae are given by:

$$\Delta = \Delta_1^1,$$

$$\Delta^2 = \Delta_1^2,$$

$$\Delta^3 = \Delta_1^3.$$

Here Δ is called the first difference, Δ^2 the second difference, and Δ^3 the third difference.

Nowadays this formula is commonly known as Newton's interpolation formula. In Europe it was first used by the British astronomer Gregory and later it was further generalized by Newton at the end of the seventeenth century.

The interpolation formula used by astronomers of the Suí and Táng periods is equivalent to using the first three terms of the above formula and regarding the third difference Δ^3 as zero. That is considering only:

$$f(w + s) = f(w) + s\Delta + \frac{s(s-1)}{2!} \Delta^2.$$

This way of considering only as far as the second differences is called the 'method of second differences interpolation'.

The first man to apply the second differences interpolation formula for calculating the position of the sun and the moon was Liú Zhuó, whom we introduced above. Liú Zhuó was a famous astronomer in the Suí Dynasty. In 600 AD he edited and compiled a new calendar called the 'Imperial Standard Calendar'. In the 'Imperial Standard Calendar' (皇極曆 , Huáng jí lì) Liú Zhuó applied the equal interval second differences method of interpolation for the computation.

Suppose at times w, $2w$,. $3w$, . . . the observed results are $f(w)$, $f(2w)$, $f(3w)$, . . . and again let $d_1 = f(2w) - f(w)$, $d_2 = f(3w) - f(2w)$ (that is, $d_1 = \Delta_1^1$ and $d_2 = \Delta_2^1$ in the differences formulae given above), then Liú Zhuó's computation used the following formula:

$$f(w + s) = f(w) + \frac{s(d_1 + d_2)}{2} + s(d_1 - d_2)$$

$$- \frac{s^2}{2} (d_1 - d_2) .$$

It is very easy to prove that this formula and the formula using the first three terms of Newton's formula are equivalent for it only requires the following verification:

$$f(w + s) = f(w) + \frac{s(d_1 + d_2)}{2}$$

$$+ \quad s(d_1 - d_2) - \frac{s^2}{2} (d_1 - d_2)$$

$$= f(w) + s \left[\frac{d_1 + d_2}{2} + \frac{2d_1 - 2d_2}{2} - \frac{s(d_1 - d_2)}{2} \right]$$

$$= f(w) + s \left[d_1 + \frac{(d_1 - d_2) - s(d_1 - d_2)}{2} \right]$$

$$= f(w) + s \left[d_1 + \frac{(s - 1)}{2} (d_2 - d_1) \right]$$

$$= f(w) + sd_1 + \frac{s(s - 1)}{2} (d_2 - d_1)$$

$$= f(w) + s\Delta + \frac{s(s - 1)}{2} \Delta^2$$

$$(\because \Delta = d_1, \Delta^2 = d_2 - d_1).$$

In the middle of the Táng Dynasty the famous astronomer Monk Yì Xíng (一行) used the unequal interval second differences interpolation formula for calculations. This method of computation was recorded in the 'Dà Yǎn Calendar' (大衍曆 , 727 AD), which was compiled and edited by Monk Yì Xíng.

Assume L_1, L_2 are two unequal time intervals and at times w, $w + L_1$, $w + (L_1 + L_2)$ the results of observations are $f(w), f(w + L_1), f(w + [L_1 + L_2])$.

Let
$$d_1 = f(w + L_1) - f(w),$$
$$d_2 = f(w + [L_1 + L_2]) - f(w + L_1)$$

then the unequal interval formula of Yì Xíng is equivalent to:

$$f(w + s) = f(w) + s \frac{d_1 + d_2}{L_1 + L_2} + s \left(\frac{\Delta_1}{L_1} - \frac{\Delta_2}{L_2} \right)$$
$$- \frac{s^2}{L_1 + L_2} \left(\frac{\Delta_1}{L_1} - \frac{\Delta_2}{L_2} \right).$$

Towards the end of the Táng Dynasty Xú Áng (徐昂) compiled and edited the 'Xuān Míng Calendar; (宣明曆 , 822 AD) and he simplified the unequal interval second differences interpolation formula of Yì Xíng to:

$$f(w + s) = f(w) + s \frac{d_1}{L_1} + \frac{sL_1}{L_1 + L_2} \left(\frac{\Delta_1}{L_1} - \frac{\Delta_2}{L_2} \right)$$
$$- \frac{s^2}{L_1 + L_2} \left(\frac{\Delta_1}{L_1} - \frac{\Delta_2}{L_2} \right).$$

When computing the position of the moon, Xú Áng used the equal interval second differences formula. This formula has the following form:

$$f(w + s) = f(w) + sd_1 + \frac{s}{2}(d_1 - d_2) - \frac{s^2}{2}(d_1 - d_2).$$

This is closer to the form of Newton's formula and can easily be simplified to

$$f(w + s) = f(w) + s\Delta + \frac{s(s - 1)}{2}\Delta^2,$$

$$(d_1 = \Delta, d_2 - d_1 = \Delta^2).$$

4.2 The *Ten Books of Mathematical Classics* (十部算經 , Shí bù suànjīng) and mathematical education during the Suí and Táng dynasties

1 The *Ten Books of Mathematical Manuals*

After close to a thousand years of development from the Hàn to the Táng dynasties, mathematics in ancient China had gradually been formed into a complete system.

In the history of mathematics in ancient China many outstanding mathematicians and many books appeared during these thousand years or so. In the *Book of Arts and Crafts* in the *Hàn history* it is recorded that there were two examples of mathematical writing; in the *Book of Arts and Crafts* in the *Suí history* this was increased to 27 titles, in the *Book of Arts and Crafts* in the *Old Táng history* there were 19 titles recorded and in the *Book of Arts and Crafts* in the *New Táng history* there were 35 titles. Amongst them the most famous is the *Commentary on the Ten Books of Mathematical Classics* commissioned by the Emperor and annotated by Lǐ Chùnfěng (李淳風) and others. The *Ten Books of Mathematical Classics* was culled from a number of mathematical writings and was finally approved as the textbook used by the Imperial Academy (國子監 , Guózǐ Jiàn) and for civil service examinations. In particular it seems appropriate to say that although all the other mathematics works are lost, the book which *was* passed down to the present is indeed a very valuable item for understanding the development of mathematics in China during more than a thousand years from the Hàn to the Táng periods. (*The Method of Interpolation* by Zǔ Chōngzhī was lost long ago.)

The *Zhōubì*, *Nine Chapters*, *Sea Island*, etc. have been discussed briefly in earlier sections. Below, we give a brief introduction to *Master Sūn's Mathematical Manual*, *Zhāng Qiūjiàn's Mathematical Manual*, and other books.

Master Sūn's Mathematical Manual (孫子算經 , Sūnzǐ Suànjīng)

It is not very clear when *Master Sūn's Mathematical Manual* became a book. In the preface to *Zhāng Qiūjiàn's Mathematical Manual* it mentions 'the problem of the rectangular storehouse of *Xiàhóu Yáng* and the washing up problem of *Master Sūn.*' The preface to *Xiàhóu Yáng's Mathematical Manual* remarks that '[*Mathematical Manual of the*] *Five Government Departments* (五曹 , *Wŭ cáo*) and *Master Sūn* reported on a lot of work'. From this we can see that *Master Sūn* was somewhat earlier than the books of Zhāng Qiūjiàn and Xiàhóu Yáng. The earliest surviving edition of *Master Sūn* is the Southern Sòng edition (slightly after 1213 AD) and is kept in the Shànghăi library. The ordinary edition is in the *Ten Mathematical Manuals* (算經十書 , *Suànjīng Shíshū*) xylographed by Kŏng Jìhán (孔繼涵) of the Qing dynasty.

Master Sūn's Mathematical Manual is divided into three volumes. The first volume treats the methods for multiplication and division, using counting rods. The second volume discusses the methods for calculating with fractions and extracting square roots. These are both very good sources for understanding calculations using counting rods in ancient China. They supplement the deficiencies in the *Nine Chapters on the Mathematical Art*. The last volume collects some difficult problems in arithmetic. For example, problems such as being given the number of heads and the number of feet for 'chickens and rabbits in the same cage': these problems still appear frequently in arithmetic textbooks in present day China.

However, in *Master Sūn's Mathematical Manual* the most famous problem is Problem 26 in the last volume. It is usually known as 'Master Sūn's problem'. The original text of this problem says: 'There are an unknown number of things. Three by three, two remain; five by five, three remain; seven by seven, two remain. How many things?' Using modern algebraic notation, it can be written as follows: There is a number N. Dividing by m_1, the remainder is r_1; dividing by m_2, the remainder is r_2; dividing by m_3, the remainder is r_3. What is N?

Writing $N \equiv r_1 \pmod{m_1}$ ['N is congruent to r_1 modulo m_1'] to mean that N divided by m_1 leaves r_1, then the problem is equivalent to solving three simultaneous congruences (that is, to finding the smallest positive integer satisfying the congruences), that is to finding N where

$$\begin{cases} N \equiv r_1 \pmod{m_1}, \\ N \equiv r_2 \pmod{m_2}, \\ N \equiv r_3 \pmod{m_3}. \end{cases}$$

The solution is not too difficult. It is only necessary to find a_1, a_2 and a_3 such that a_1 divides exactly by m_2 and m_3 but has remainder 1 when divided by m_1; a_2 divides exactly by m_1 and m_3 but has remainder 1 when divided by m_2; and a_3 divides exactly by m_1 and m_2 but leaves remainder 1 when divided by m_3. Then the solution of the simultaneous congruences is $a_1 r_1 + a_2 r_2 + a_3 r_3$ and one then obtains the smallest positive integral solution by successive subtraction of the least common multiple of m_1, m_2 and m_3.

'Master Sūn's problem' circulated quite widely among the ordinary people of China and bears other names such as 'the Emperor of Qín (秦) counting his soldiers secretly', 'The field commander Hán Xin (韓信) counting soldiers', 'The method of flights [of arrows]', and 'The Guĭ Gŭ (鬼谷) calculation'. A manuscript of the Sòng Dynasty has the solution of the problem in a four-line verse:

Three-year child at seventy makes a change,
From five, twenty-one remaining is strange,
Meet again seven times at Shàng Yuán,
Qing Ming's cold food day is the answer then.

Shàng yuán (上元) means the fifteenth day: it comes from an ancient meaning of 'the first fifteen days of the first lunar month'. Qing míng (清明) 'clear and bright' is the ancient name for one hundred and six days after the winter solstice. The 'cold food eating day' (寒食 , hán shí) is the day before qing míng and so means (day) 105. In this rhyme the three numbers, 70, 21 and 15 are the right choices for a_1, a_2, and a_3, and 105 is the least common multiple of m_1, m_2, and m_3.

'Master Sūn's problem' is not only a very interesting problem in its own right, it is also related to the computation of the calendar in ancient China. Suppose that N years ago at midnight at the winter solstice the sun, the moon and the five known planets were all at the same bearing. This can be regarded as a common starting point for the position of the sun, the moon, and the five known planets. Now the sun, the moon, and the five known planets all have different periods of rotation, so if observed N years later at the Qth hour of the Pth day of the Mth month, their positions in their trajectories are different. Let m_1, m_2, m_3, . . . be the periods of the motions of the sun, the moon, and the five known planets, respectively. Use these to divide N years M months P days Q hours, giving remainders r_1, r_2, r_3, . . . then these indicate the displacement of the sun, the moon, and the five known planets from their starting points to their present positions. Conversely, by knowing m_1, m_2, m_3, . . . and r_1, r_2, r_3, . . . one can find the total number of years, N, as in the solution of Master Sūn's problem. When this sort of calculation was used to find the year in ancient times it was called the Shàng Yuán (上元 , previous origin) method and N was called the accumulated years from the Shàng Yuán. Even now it is still not clear when they began to investigate the calculation of the accumulated years from the Shàng Yuán (previous origin). However, in the Dà Míng calendar of Zŭ Chōngzhī (462 AD) the method of calculation was very complicated. There he considers 11 parameters. From the mathematical point of view this means solving 11 simultaneous congruences for N. In the computation of the calendar the periods (m_1, m_2, . . .) of the heavenly bodies are not integers, so a_1, a_2, . . . are not easy to determine. Unfortunately the methods of calculation used by the astronomers of those days have not been passed down to the present and we have to wait till the 13th century AD before there is a systematic description in the mathematician Qín Jiǔsháo's book *Mathematical treatise in Nine Sections* (數書九章 , *Shùshū jiǔzhāng*) (1257 AD) in the Sòng Dynasty (details later).

Mathematical Manual of the Five Government Departments (五曹算經 , Wǔcáo suànjīng), Arithmetic in the Five Classics (五經算術 , *Wǔjīng suànshù*), and Memoir on some Traditions of Mathematical Art (數術記遺 , *Shùshù jìyì*)

These three books are credited to Zhēn Luán (甄鸞) whose other name was Shūzūn (叔遵); he was either author or commentator. Southern Sòng editions of the last two works are preserved in Běijīng (Peking) University Library. The Southern Sòng edition of the *Mathematical Manual of the Five Government Departments* is lost and the presently surviving edition is copied from the *Great Encyclopaedia of the Yǒng-lè reign period* (*Yǒng lè dàdiǎn*, 永樂大典). The common edition is still the one published by Kǒng Jìhán (孔繼涵) of the Qīng Dynasty in the *Ten Mathematical Manuals* (算經十書 , *Suànjīng shíshū*). Zhēn Luán (甄鸞) lived in the period of the North and South Dynasties (sixth century AD) in northern Zhōu. He was a Buddhist, skilled in astronomy and the making of calendars. He compiled the 'Tiān Hé calendar' (天和曆), which was officially adopted in the first year of the Tiān Hé reign in northern Zhōu (566 AD).

The *Mathematical Manual of the Five Government Departments* was compiled especially for the various categories of officials. It may be described as a textbook on applied arithmetic. In the *Book of Arts and Crafts* in the *Old History of the Táng Dynasty* it says that the book was written by Zhēn Luán. The existing edition does not mention who the author was; it simply says it was 'commented on by Lǐ Chūnfēng (李淳風) in the Táng Dynasty and others'.

It is called the *Five Government Departments* to indicate the departments (曹 , cáo) for farmland, the army, customs, warehouses, and finance. There are five chapters in the book: one chapter for each department.

The contents of the chapter for the farmland department concern the calculations of areas of cultivated fields. Besides the calculations of rectangular, triangular, trapezoidal, and circular areas, which are the same as in the *Nine Chapters*, there are also other areas that are calculated by approximate methods. The second chapter for army officials concerns some problems on army equipment and the transport of supplies. The methods of computation in this chapter appear to be within the scope of the *Nine Chapters* but it records some of the earliest systematic applications of mathematics to military matters in China. The third chapter for customs officials has problems on the exchange of trade. The fourth chapter for warehouse officials is about problems on the taxation of foodstuffs and calculations on the capacity of warehouses. Chapter five is for the finance officials and contains problems on the silk trade and the management of government income and expen-

4	9	2
3	5	7
8	1	6

Fig. 4.1 Magic square.

diture. These three chapters use calculation methods which are still within the ambit of the *Nine Chapters*. The methods are multiplication, division, and proportions.

In the *Arithmetic in the Five Classics* Zhēn Luán discusses in detail statements about mathematics in the *Ancient History* (尚書 *Shàng Shū*), the *Classic of Poetry* (詩經 , *Shī Jīng*), *Book of Changes* (周易 , *Zhōu Yì*), the *Zhou Rites* (周禮 , *Zhōu Lǐ*), the *Record of Rites* (禮記 , *Lǐ Jì*), and he explains some of the writings of mathematicians in the Hàn period. As far as mathematics is concerned there are not very many outstanding items nor can one find much help here for understanding and studying the classics.

The book *Memoir on some Traditions of Mathematical Art* (數術記遺 , *Shùshù jì yí*) was not originally counted as one of the *Ten Mathematical Manuals* but during the thirteenth century the people of the Sòng period who were republishing the *Ten Mathematical Manuals* found that the *Method of Interpolation* had long been lost. They could not find a copy but unexpectedly they chanced on the *Memoir on some Traditions of Mathematical Art* and then included it. Thus the existing edition of the *Ten Books of Mathematical Classics* came about.

The *Memoir* uses many Buddhist, Taoist, and mystical phrases in its explanations. This means the book contains a lot of mysticism and occultism. The contents are not easy to understand. The book mentions other methods of manipulating numbers besides the counting rods, but all of these are impractical. The book introduces the 'row and column diagram', which contains three rows and columns as in Fig. 4.1, and calls it the 'nine houses computation'. The phrase 'nine houses' originated in the Hàn period in the fortune-telling school.

In the tenth century AD at the beginning of the Sòng Dynasty, some people connected the 'nine houses' diagram and the Luò-chū-shū (洛出書) in the Xìcí chapter in the Yì Jīng (易 · 繫辭 , the *Great Appendix to the Book of Changes*) and called it Luò-shū (洛書). They contended that the Luò-shū is the very basis of the origin of mathematics. That is unbelievable.

The 'row and column diagram' is what is today known in mathematics as a 'magic square'. As in Fig. 4.1 the sums of the numbers in each of the three rows, three columns, and in the main diagonals are all equal to 15. The method of formation is no secret at all (see Fig. 4.2(a)). First draw a diamond, divide it into nine small squares and enter 1, 2, . . ., 9 sequentially

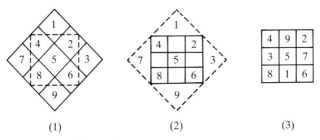

Fig. 4.2 The formation of the magic square.

into these. Next draw an upright square (dotted line) over the diamond and again divide it into nine small squares. This has four blank squares as in Fig. 4.2(b). Finally put the remaining four numbers in the blank spaces after interchanging the upper and lower and left and right positions. This yields a three-row magic square as in Fig. 4.2(c). The sums of the numbers along each row, each column and both main diagonals are all equal to 15.

There are more detailed discussions of these arrangements of numbers in the mathematical works of Yáng Huí (楊輝) in the 13th century in the period of the Southern Sòng dynasty; from then on they were called 'row and column diagrams' ['magic squares']. Yáng Huí also described the formation of the magic square by the above method of interchanging. In fact, all the magic squares using successive numbers as the magic numbers can be formed in this way. (See also Ho Peng-Yoke 1973, and Lam Lay-Yong 1977.)

Xiàhóu Yáng's Mathematical Manual (夏侯陽算經 , *Xiàhóu Yáng suànjīng*)

In the preface to *Zhāng Qiūjiàn's Mathematical Classic* it mentions 'Xiàhóu Yáng's square storehouse', so we can immediately conclude that *Xiàhóu Yáng's Mathematical Manual* became a book before *Zhāng Qiūjiàn's Mathematical Manual*. It is a work from about the fourth century AD.

However, the existing edition of the *Xiàhóu Yáng's Mathematical Manual* includes a lot of material that could only have appeared after the fourth century AD. Indeed, it even contains some of the rules and regulations of the Táng Dynasty of the eighth century AD. For example, in the book there are problems on the calculation of the two taxes on rice and two taxes on household wealth, but the two-tax system was only made law at the beginning of the reign of Emperor Dài Zōng (代宗) of Táng (reigned 763–779 AD). (This 'two-tax' system consisted of two levies made in summer and autumn. It replaced a system of complex taxes and surcharges.) Further, there are problems using 'government loans', 'legislation for warehouses', and 'legislation for corvée labour'. These are all laws from the Táng Dynasty. There are only two possibilities. One possibility is that the *Xiàhóu Yáng's Mathematical Manual* was lost a long time ago and the present edition is another book compiled and written by his successors and it is only because the first sentence of the book uses 'Xiàhóu Yáng said . . ., that the whole book was called *Xiàhóu Yáng's Mathematical Manual*. The other is that, in the process of transmission, the original book *Xiàhóu Yáng's Mathematical Manual* was successively copied and that, in the process of this continued copying, the copyists added problems from their own periods so that the existing edition of the *Xiàhóu Yáng's Mathematical Manual* acquired a lot of material from the fourth to the eighth centuries and thus grew in content. After reaching the Sòng period this book was published in the *Ten Mathematical Manuals*; there were no further changes and it survives to the present day.

The Southern Sòng edition of *Xiàhóu Yáng's Mathematical Manual* has long been lost and the most complete surviving edition is the one copied at the beginning of the Qing Dynasty from the Sòng book. It is now in the Běijīng (Peking) Imperial Palace Museum. A common edition is that in the *Ten Mathematical Manuals* xylographed by Kǒng Jìhán.

In the existing edition of *Xiàhóu Yáng's Mathematical Manual* there are three chapters altogether with a total of 83 problems. Most of the problems are concerned with calculations on various aspects of practical day-to-day life in the society of those days. Some of the problems are similar to ones in *Master Sūn's Mathematical Manual* and other books.

What is worth pointing out about *Xiàhóu Yáng's Mathematical Manual* is that there is a clear inclination towards improving the system of computation using counting rods. This tendency started at the end of the Táng Dynasty, continued through the Five Dynasties and Ten Kingdoms and right through the Sòng and Yuān Dynasties without a stop, until finally in the middle of the fourteenth century a new calculation device was produced. This was the bead abacus (details below).

Zhāng Qiujiàn's Mathematical Manual

The period of *Zhāng Qiujiàn's Mathematical Manual* is approximately just after *Master Sūn's Mathematical Manual* and *Xiàhóu Yáng's Mathematical Manual*. Problem 13 in the middle chapter presents a problem of allocating taxes according to nine classes of people. In the *Book of Trade: Wèi history* (魏書 · 食貨志 , *Wèi shū shí huò zhi*) it is recorded that from 466–485 AD the Tuòbá (拓跋) royal house of Wèi promulgated the taxation law 'according to the wealth, rent, and transport of the people, in the three classes, nine divisions'. So we can conclude that *Zhāng Qiujiàn's Mathematical Manual* is a book written about the middle of the fifth century AD in the North and South Dynasties period in China.

The earliest surviving edition is the Southern Sòng woodblock printing (shortly after 1213 AD), which is kept in the Shànghǎi library. The latter part of the second volume of this copy is incomplete and the first few problems of the third (last) volume are lost. The present edition has three chapters and a total of 92 problems. The most common edition is the one published by Kŏng Jíhán of the Qīng dynasty in the *Ten Mathematical Manuals*. A Russian translation has been made by Berezkina (1969).

This book paid particular attention to calculations with fractions. In the preface it says: 'All who learn to calculate are not afraid of the difficulties in multiplication and division but have difficulty in finding common denominators'. The book contains a lot of problems on the application of fractions.

Zhāng Qiujiàn's Mathematical Manual also contains several problems concerning calculations with sequences and series. The book gives some formulae. Using modern algebraic notation these can be written as follows:

1. Knowing the first term a and last term ℓ and that there are n terms, then the sum s of the n terms is:

$$s = \frac{1}{2}(a + \ell)n, \quad \text{(Chapter 1, Problem 23)}.$$

2. Given a, n and s find the common difference d:

$$d = \left[\frac{2s}{n} - 2a\right] \bigg/ (n - 1), \quad \text{(Chapter 1, Problem 22)},$$

There are several others.

The last problem in the last chapter in *Zhāng Qiujiàn's Mathematical Manual* is known as the 'hundred fowls problem'. The original problem is 'One rooster is worth five copper cash; one hen is worth three copper cash; three young chicks are worth one copper cash. Buying 100 fowls with 100 cash, how many roosters, hens and chicks?' (A copper cash was a standard Chinese coin.) This is equivalent to solving the equations

$$\begin{cases} x + y + z = 100, \\ 5x + 3y + \dfrac{1}{3}z = 100. \end{cases}$$

There are three sets of answers in the book: (4, 18, 78), (8, 11, 81), and (12, 4, 84). The answers are correct. How were these answers obtained? In the original text it simply says: 'Roosters increase four each time [that is from four to eight to 12], hens decrease by seven each time, chicks increase three each time'. The book does not explain how the adding four, subtracting seven and adding three was found, nor how the first set of answers was obtained. Considering the form of the statement it is possible that the answer was obtained by trial and error, each time adding four roosters, subtracting seven hens and adding three chicks. That makes the number of fowls and the amount of money remain unchanged. It is possible that the first set of solutions was obtained as follows: Let $x = 0$ and solve

$$\begin{cases} y + z = 100, \\ 3y + \dfrac{1}{3}z = 100, \end{cases}$$

getting $y = 25$ and $z = 75$. Starting from $x = 0$, $y = 25$, $z = 75$ and every time adding 4 to x, subtracting 7 from y and adding 3 to z the sets of three solutions can be obtained successively.

The 'hundred fowls problem' was also recorded in a lot of mathematical works of later periods such as Zhēn Luán's *Memoir on some Traditions of Mathematical Art*, Yáng Huī's *Continuation of Ancient Mathematical Methods for Elucidating the Strange (Properties of Numbers)* (續古摘奇算法 , *Xùgŭ zhāiqí suànfă*, 1275 AD) and Chéng Dàwèi's (程大位) *Systematic Treatise on Arithmetic* (算法統宗 , *Suàn fă tŏng zōng*, 1592 AD) as well as other works of the Míng and Qīng Dynasties. One can also find similar problems in mathematical books written in ancient India or in the Islamic countries of the Middle Ages. For example the Indian mathematician Bhaskara of the 12th century AD has a problem in his work which is just the same as the 'hundred fowls problem' in *Zhāng Qiujiàn's Mathematical Manual*. The 15th century Arab mathematician al-Kasi also introduced a problem very similar to the hundred fowls problem.

In *Zhāng Qiujiàn's Mathematical Manual* there are also two problems on the solution of quadratic equations. These two problems are equivalent to solving:

$$x^2 + 68\frac{3}{5}x = 514\frac{32}{45} \times 2 = 1029\frac{19}{45},$$

(surviving edition, last problem of the middle chapter) and

$$x^2 + 15x = 594.$$

(surviving edition, Problem 9 of last chapter).

In *Zhāng Qiujiàn's Mathematical Manual* there are detailed procedures and explanations for the extraction of square and cube roots using counting rods, but as for the solution of quadratic equations in general — the 'corollary to the square root method' (see above, Chapter 2) — there is no explanation at all. The solution of the last problem in the book ends with 'take the field and add the number of paces, doubling it (as the given term) and using the number of paces in the hypotenuse as the derived element'; the following pages are lost and what follows is not known.

Continuation of Ancient Mathematics (緝古算經 , *Xùgŭ suànjīng*)

Continuation of Ancient Mathematics is a work from the early Táng Dynasty (i.e. early seventh century AD); its author was Wáng Xiàotōng (王孝通). Neither the actual date of the compilation of the book nor the birth and death dates of the author are now known. Wáng Xiàotōng was not only a mathematician but also an astronomer in those days. This was recorded in the *Book of the Calendar: New History of the Táng Dynasty* (新唐書 · 曆志 , *Xīntáng shū · Lì zhì*) in the sixth (623 AD) and ninth years (626 AD) of the Wŭ Dé

(武德) reign. Wáng Xiàotōng was a scholar in the computation of the calendar and he criticized the 'Wù Yín calendar' (戊寅曆) (618 AD) edited by Fù Rénjūn (傅仁均). In the 'Address at the presentation of the *Continuation of Ancient Mathematics*' Wáng Xiàotōng said to the Emperor:

'Your loyal subject was brought up in the nobility, learned calculation when young, shed his foolishness, now his hair is almost grey and he comes to bow before you . . . Your Majesty was generous in reappointing him [he had been the official under the Suí dynasty and was then appointed under the Táng], in using your subject as State Astronomer (Tài shi chéng, 太史丞) and in these few years when ordered to study the Fù Rénjūn calendar he has made corrections to errors in the calculations in more than thirty places and these are now being incorporated in the calendar by the astronomers.

In this passage the events clearly took place in the sixth and ninth years of the Wǔ Dé reign. According to the above passages from the presentation address, the work *Continuation of Ancient Mathematics* must have been completed after the ninth year of the Wǔ Dé reign (after 626 AD), so it can be regarded as a work written in the first half of the seventh century AD.

There are 22 problems in the book. In addition to Problem 1, which is on the calculation of the position of the moon and so is connected with astronomy and the computation of calendars, Problems 2–5 are on the construction of platforms, embankments, river courses, and so on. Problems 6–14 are problems about the construction and repair of various types of food warehouses and storage caves. Problems 15–20 are problems connected with right-angled triangles, so-called Gōugǔ problems. In the surviving edition problems, 17, 18, 19, and 20 are incomplete, only the first 16 problems being complete. The original Southern Sòng edition of the *Continuation of Ancient Mathematics* has long been lost, the present day edition being based on the copy of the Southern Sòng edition made by the Míng Dynasty bibliophile Máo (毛) of Jí Gǔ (汲古) pavilion private library. This hand-copied edition is stored in the Běijīng Imperial Palace Museum. It is a book in the Imperial Collection of Heavenly Bestowed Classics (天祿琳瑯 , Tiánlù línláng). All the other versions are based on this edition.

The most important items in the book concern problems on the construction of embankments whose ends are not of the same width and whose heights are different. In the 'Address at the presentation of *Continuation of Ancient Mathematics*', Wáng Xiàotōng himself said:

. . . People in our day say of Zǔ Gěng's 'Method of Interpolation' that it is superb but they did not detect that the methods for [calculating the] construction of fortifications are all incorrect and for the problems on square and rectangular frusta the treatment is incomplete. Your subject has now produced new methods which go further . . . Looking through the 'Construction consultations' chapter in the *Nine Chapters on the Mathematical Art* one finds methods for the 'construction of bodies on level ground' but wide tops, narrow bottoms, high fronts or low backs are not discussed in any of the Classics. This led to the present generation being unable to understand the theory of depth, for in simple situations as compared with irregular ones, people only use the circle, cylinder, and square. How does one deal with the irregular ones? Your subject thought day and night, studying all the Classics, afraid that any day his eyes would close forever and the future not be seen; he progressed from level surfaces and extended that to narrow and sloping objects [yielding] twenty methods in all in the *Continuation of Ancient Mathematics*. Please request capable mathematicians to examine the worth of the reasoning. Your subject will give a thousand gold coins for each word rejected. . . .

The problems of constructions involving wide topped, narrow based, high fronted, low backed irregular shapes were indeed the most outstanding part of the book.

Among them Problem 3 may be taken as a model example. Using modern mathematical language the problem can be presented as follows:

An embankment is being constructed with a trapezoidal cross-section. At the western end the top and bottom differ in width by 68.2 [Chinese] feet and at the eastern end they differ by 6.2 feet. The height of the eastern end of the embankment is 3.1 feet less than the western end and the width of the top of the embankment at the eastern end is 4.9 feet more than the height, while the total length of the embankment is 476.9 feet more than the height of the eastern end. A, B, C, D are four districts sending 6724, 16 677, 19 448, and 12 781 men, respectively, for the construction of the embankment. Each day each man can dig 9 shí 9 dǒu 2 shēng [= 0.992 m³.] of earth or can build up 11.4$\frac{6}{13}$ cubic feet of embankment, every cubic foot dug being 8 dǒu of earth. Each man can carry 2 dǒu 4 shēng 8 gě and can move 192 steps on level ground 62 times. At present they have to cross a hill and a stream to get the earth, a flat distance of 11 paces, an incline of 30 paces on the hill and the width of the stream is 12 paces. Carrying heavy things up the hill three steps are equivalent to four on level ground, going downhill six steps are equivalent to five on level ground and crossing the stream one step is equivalent to two on level ground while in the course of construction delays for various reasons add the equivalent of one step to every 10, and the job of removing earth is equivalent to carrying weight for 14 paces. The four districts doing the construction complete the job in one day. The four districts successively distribute their men in four sections from east to west. What is the length of the top and the bottom, the height and the width, in feet, of each section of the construction? Further, how much can one man construct in one day doing all the digging, transporting, and mounding up? Again, what are the widths of the top and bottom and the height of the eastern and western ends in feet?

It is obvious that this question is very complicated; 25 calculations are required.

The shape of the embankment is shown in Fig. 4.3. The top width is constant but the bottom widths of the two ends are different and also one end is higher than the other.

First of all, from the information given in the problem, calculate the amount of embankment completed by each man doing the digging, transporting, and building up on his own. Then multiply by the numbers of people

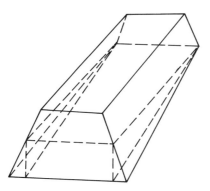

Fig. 4.3 Non-regular embankment.

from the four districts to get the volume of the whole construction. This is the volume V of the whole embankment. Further, Wáng Xiàotōng divided the whole embankment into four parts as in the figure. The top part is just an ordinary embankment (the heights of the two ends are the same and so are the top and bottom widths); the middle part of the bottom is a prism (diagonal section of a parallelepiped) and the two sides are tetrahedra (鼈臑 , biē'nào See Chapter 3, p. 71). Let the height of the lower end be x, then using the conditions on the end measurements and the previously calculated volume V one can write out a cubic equation:

$$x^3 + ax^2 + bx = A.$$

Solving this, one gets x and thence the widths of the top and bottom, the heights, and the length of the embankment, etc.

Wáng Xiàotōng did not use the algebraic method similar to saying 'Let the height be x' but formed his equation from the shape of the geometric object. Wáng Xiàotōng called the coefficients of the above-mentioned cubic shi (A), fāng (b), lián (廉) (a), and he called the general method of solving this type of cubic equation 'the corollary to taking cube roots'. This is the 'cube root derived method'. The 'corollary to taking cube roots' evolved directly from the method of taking cube roots and its invention was probably quite early. This method is an extension of the method of extracting square roots and its corollary (see above, Chapter 2, p. 50). However, according to the available material the presently circulating edition of *Continuation of Ancient Mathematics* by Wáng Xiàotōng is the first work recording the method of the 'corollary to taking cube roots' in the mathematics of ancient China.

To find the length of the sections when the work is distributed among the men from the districts A, B, C, D given the volume of the embankment again requires solving a cubic equation.

In the 20 problems of the *Continuation of Ancient Mathematics* there are

28 cubic equations. It is worth pointing out that the coefficients in all the equations given by Wáng Xiàotōng are positive numbers, their roots are only positive ones and only one root is given. In the period of the Sòng and Yuán Dynasties, from the 11th–13th centuries AD there were distinct improvements in the methods of solving equations in the mathematics of ancient China. Not only were equations of arbitrary degree solved but also coefficients were not restricted to being positive, and further there was a complete classification system for equations (details below).

2 Mathematical education in the time of the Suí and Táng Dynasties

The historical record of mathematical education in ancient China has been handed down to the present from very early times (details may be found in previous chapters).

After reaching the Suí Dynasty, when a unified country had emerged, various types of uniform rules and regulations were promulgated as laws for the whole country. As far as the education system is concerned in schools controlled by the government — the Imperial Academy (國子監 , Guózǐ Jiàn) (which is equivalent to the present-day State Universities) — mathematics was taught among other subjects. In the report in the *Book of Officials* in the *History of the Suí Dynasty* (隋書 · 百官志 , Suí shū · Bǎi guān zhì) it says: 'The Chancellor of the Imperial College . . . was in charge of the following subjects: Propriety (統國子 , Tǒng guózǐ), the Great Learning (太學 , Tài xué), Four Ways (四門 , Sì mén), history, and mathematics, and each subject had positions for scholars, teaching assistants, and students'. According to the commentary in the *Book of Officials: Suí history* at that time for mathematics in the Imperial Academy there were two scholars, two teaching assistants and 80 students.

As in the case of the Hàn Dynasty, which had adopted most of the rules and regulations of the Qín Dynasty, the Táng Dynasty had adopted most of the rules and regulations of the Suí Dynasty. Also the mathematical education under the Táng was taken over from that of the Suí and so evolved from it. According to the available material, at the beginning of the seventh century AD in the second year of the Zhēn Guān (貞觀) reign of the Táng (628 AD), the subject of mathematics was established in the Imperial Academy.

However, other sources claim it was not until the middle of the seventh century, in the first year of the Xiǎn Qìng (顯慶) reign (656 AD) that the subject of mathematics was established.

In the several centuries of the Táng Dynasty the subject of mathematics was sometimes introduced, sometimes cancelled, sometimes belonged to the Imperial Academy, sometimes to the Institute of Astronomy or the Institute of Records.

After the re-establishment of the subject of mathematics in the third year of the Xiǎn Qìng reign (658 AD) it was said that 'Since mathematics . . .

leads only to trivial matters and everyone specializes in their own way, it distorts the facts and it is therefore decreed that it shall be abolished'. Further, the scholars and others in mathematics were transferred to the Institute of Astronomy. The second year of the Lóng Shuò (龍朔) reign (662 AD) we have: 'Re-establish . . . the officials in mathematics, one person', and the third year of the same reign (663 AD): 'Mathematics again transferred to the Institute of Records'.

From the material available it appears that throughout the Táng Dynasty there were great changes in the subject of mathematics in the Imperial Academy in the number of professors, teaching assistants, students, and so on. In the Táng encyclopaedia (唐六典, Táng liù diǎn) it is recorded that there were 30 students. According to records in the *History of the Administrative Statutes of the Táng Dynasty* (唐會要, *Táng huì yào*) in the second year of the Yuán Hé (元和) reign (807 AD) the Vice-Chancellor reported: 'Students in Universities in the two capitals [長安、洛陽, Cháng'ān and Luòyáng], . . . in the Institute of Mathematics, 10 people.' In the twelfth month of the same year the Emperor decreed: 'In the Eastern Capital [Luòyáng], placement of students in the Imperial Academy . . . the Institute of Mathematics, two people.' At that time the number of people was very small indeed.

Concerning the sources of students the Táng encyclopaedia (唐六典, Táng liù diǎn) records: 'The scholars in mathematics are in charge of teaching the children of the military and civil officials from class 8 down and those of commoners.' That is to say, the sources of mathematics students in the Imperial Academy were the children of the lower ranks of officials and of commoners.

The textbooks used in those days were uniformly prescribed by the government. It is reported in the *Records of Official Books* (册府元龜, Cè fǔ yuán guī) in the first year of the Xiǎn Qíng reign (656 AD): '. . . Yú Zhìníng (于志寧), and others asked the Emperor to decree that the *Ten Mathematical Manuals* should be used throughout the country.' It is also reported in the New and Old Histories of the Táng Dynasty:

'At the beginning of Táng . . . in his report to the Emperor, Wáng Sibiàn (王思辯) said: A lot of the arguments in the *Mathematical Manual of the Five Government Departments* and others in the *Ten Mathematical Manuals* are contradictory. Li Chūngfēng (李淳風) together with the scholar in mathematics in the Imperial Academy Liáng Shù (梁述), the teaching assistant Wáng Zhēnrú (王眞儒) and others were ordered to write commentaries on the *Ten Mathematical Manuals* such as the *Mathematical Manual of the Five Government Departments* and *Master Sūn's Mathematical Manual*. After the completion of these books Emperor Gāozōng (高宗) of Táng decreed their use throughout the State Institutes.

This is the first time in Chinese history that an Emperor decreed a set of textbooks for mathematics be used. These are the *Ten Mathematical Manuals* which were introduced briefly in the paragraphs above.

According to the report in the Táng encyclopaedia, to complete a course in mathematics required seven years. The syllabus of the complete course was as follows: 30 students are divided into two streams: 15 persons learn the *Nine Chapters, Sea Island, Master Sūn, Five Government Departments, Arithmetic in the Five Classics, Zhāng Qiujiàn, Xiàhóu Yáng* and *Zhōubi* and the other 15 learn the more complicated *Method of Interpolation* and *Continuation of Ancient Mathematics*. The allocation of time was *Master Sūn* and *Five Government Departments*: one year; *Nine Chapters* and *Sea Island*: three years; *Zhāng Qiujiàn* and *Xiàhóu Yáng*: one year each; *Zhōubi* and *Arithmetic in the Five Classics*: one year — a total of seven years. People in the other stream had four years to learn *Method of Interpolation*, three years to learn the *Continuation of Ancient Mathematics*; it also took seven years to graduation.

During the Táng Dynasty there was another system of civil examinations besides the above-mentioned education system for mathematics. In the civil examination there was a subject called 'Understanding Mathematics' (明算 , ming suàn). All those people who did not go through the mathematics course in the Imperial Academy were eligible to sit for the examination. In the examination there were three problems on the *Nine Chapters*, one question each on *Sea Island* and *Master Sūn* and seven others; the other people were examined by seven questions on the *Method of Interpolation* and three questions on the *Continuation of Ancient Mathematics*. Six correct answers out of ten counted as a pass. In addition they were also required to do a recitation examination on *The Memoir on some Traditions of Mathematical Art* and *Three Kinds of Calculation* (三等數 , Sān děng shù), where '9 out of 10 is a pass'. What is called 'recitation' consisted of one or two sentences being picked at random from the text and the student being asked to continue reciting from the given passage. In fact this sort of examination required memorizing passages in the *Ten Mathematical Manuals*. This sort of examination did not emphasize originality but recitation. This sort of examination cannot advance the development of science much. The candidates who passed the whole examination entered 'the lower division of class nine' — that is, the lowest class in the official ranks.

3 Mathematical interaction with the outside world during the period of the Sui and Táng Dynasties

Right from the beginning there was interaction between China and the Central Asian countries and between China and India. In the period from the North and South Dynasties to the Sui and Táng Dynasties (316–907 AD), following the penetration of Buddhist teaching, cultural exchange between China and India developed greatly. Astronomy, medicine, music, and the arts were included among the main elements of this exchange. The exchange of mathematical knowledge in both directions was also involved in the course of this sort of cultural exchange.

According to the *Book Reports in the History of the Sui Dynasty* in those days there was a total of three Chinese translations of works on Indian astronomy and Indian mathematics: *The Method of Calculation of the Brahma School* (婆羅門算法 , *Póluómén suàn fǎ)* in three volumes, *Calculation of the Calendar, Brahma School* (婆羅門陰陽算曆 , *Póluómén yīn yáng suàn lì*) in one volume and *Mathematical Manuals of the Brahma School* (婆羅門算經 , *Póluómén suànjīng*) in three volumes. These are the earliest records of the translation of foreign mathematical works into Chinese. Unfortunately all these books were lost long ago and there is no way to ascertain their contents.

After the beginning of the Táng Dynasty there were frequently quite a number of Indian astronomers taking up official positions in the State Observatory and among them the most famous was Levensita, who once held the position of State Astronomer. In the sixth year of the Kāi Yuán reign (開元 , 718 AD) he translated the Hindu *Catching Nines Calendar* (九執曆 , *Jiǔ zhi lì*) into Chinese. This method of computing the calendar by Levensita was reported in chapter 104 of the *Kāi Yuán reign-period Treatise on Astrology (and Astronomy)* (開元占經 , *Kāi yuán zhānjīng*) which has a total of 120 chapters and survives to the present. In the *Catching Nines Calendar* the following items of mathematical knowlege are introduced:

1. The measuring of arcs. It introduced the Greek method of mensuration by dividing the whole circle into 360 degrees and each degree into 60 minutes, which is the method of mensuration still commonly used today. The ancient practice of Chinese astronomy was to divide the circle into $365\frac{1}{4}$ degrees according to the number of days in a year. The Hindu mathematics method of dividing into 360 degrees and then into 60 subdivisions did not arouse the general interest of Chinese mathematicians.

2. The table of sines. In the *Catching Nines Calendar* a mathematical table for the trigonometrical sine function is presented. The interval in the table is $3°45'$ from $0°$ to $90°$ in exactly 24 steps, giving 24 values of the sine function. Using modern notation this table can be written as in Table 4.1.

Table 4.1

Interval	Degrees	$3438 \sin x$	Difference
1	3° 45′	225	
2	7° 30′	449	224
3	11° 15′	671	222
4	15°	890	219
.	.	.	.
.	.	.	.
.	.	.	.
12	90°	3438	7

Although research into trigonometry has a very special place indeed in Hindu mathematics, it did not arouse the interest of the Chinese mathematicians.

3. Hindu numerals. From the records in the *Kāi Yuán Reign-Period Treatise on Astrology (and Astronomy)* we know of the introduction of the Hindu decimal system numerals (this is the source of the present-day so-called Arabic numerals). However, mathematicians in China did not adopt them.

In the *Book of the Calendar: New History of the Táng Dynasty* it says: 'The computation of [the Catching Nines Calendar] used pen and paper and not counting rods. The method is complicated and fragmented and you are lucky if you get it right. It cannot be adopted as a method. The writing of the numbers is very strange and from first impressions it is difficult to argue for it.' That is to say that the knowledge of mathematics and astronomy introduced from India did not have a great influence on Chinese mathematics and astronomy.

The mathematics that passed from India to China and did affect Chinese mathematics was the recording of large and small numbers. Methods of recording large and small numbers had been introduced very early in the Buddhist classics: by the Sòng and Yuán period some of these names had been adopted in Chinese writings. For example when Zhū Shìjié (朱世杰) of the Yuán Dynasty in his work *Introduction to Mathematical Studies* (算學啓蒙 , *Suàn xué qi méng* — 1299 AD) introduced the recording of large numbers, the names he used included 'supreme' (極 , jí) for 10^{88}, 'the sand of the Ganges' (恒河沙 , héng hé shā) for 10^{96}, two terms with Buddhist associations (阿僧祇 , ā sēng qi) for 10^{104} and (那由他 , nà yóu tā) for 10^{112} as well as 'unimaginable' for 10^{120} (不可思議 , bù kě si yi) and 'non-countable number' for 10^{128} (無量數 , wú liàng shù) and for recording small numbers the names he used included 'a second' (須臾 , xū yú), 'twinkling of an eye' (瞬息 , shùn xi), 'flick of fingers' (彈指 , tán zhi), 'void' (虛 , xū), 'empty' (空 , kōng), 'clear' (清 , qing), and 'clean' (淨 , jing) etc. All these are quoted from the translation of Buddhist classics concerning the method of recording large and small numbers into Chinese.

In the time of the Suí and Táng Dynasties there were outstanding developments in the cultural exchange between China, Japan, and Korea. The mathematical works and the mathematical education system passed into Korea and Japan. They adopted a mathematical education system similar to the one in the Imperial Academy of the Táng Dynasty in Korea and Japan. They also adopted the *Zhōubi*, the *Nine Chapters* and altogether 10 books as textbooks.

5
The zenith of the development of mathematics during the Sòng and Yuán Dynasties

5.1 A brief description of mathematics in the Sòng and Yuán period (960–1368 AD)

Over a period of a thousand or so years, from the Hàn to the Táng dynasty, mathematics developed into a complete system whose basic content was based on the *Ten Books of Mathematical Classics*. From the tenth to the fourteenth centuries AD, in the time of the Sòng and Yuán Dynasties, there were again new developments.

In the seventh year of the Yuán Fēng (元豐) reign of the Northern Sòng Dynasty (1084 AD), the technique of wood-block printing was highly developed. The Secretariat which was in charge of old and new books, of recording the main events in the country, and of supervising the work on astronomy and the computation of the calendar, had the *Nine Chapters on the Mathematical Art* and various other books of the Hàn to Táng periods printed. The government ordered these books to be used as textbooks in schools and universities. This is the earliest set of printed textbooks recorded in Chinese history. Although the original editions have not survived to the present, the reprints of these books in the Southern Sòng period are still available in their original form (cf. Section 8.2.2).

During the Northern Sòng period (960–1127 AD) mathematics was also established as a subject in the Imperial Academy. But sometimes it was brought in and at other times it was cancelled, so it had no continuity of development. For example, in the seventh year of the Yuán Fēng reign (1084 AD) mathematics was ordered to be reintroduced and it was also planned to build and repair the Institute of Mathematics' buildings. However in the first year of the Yuán Yòu (元祐) reign (1086 AD), because 'after the introduction [of mathematics], staffing the school and setting up the civil service examinations [in mathematics] there was a general feeling that it was extravagant and did not really help in the running of the country' the subject was discontinued. In the third year of the Chóng Níng (崇寧) reign (1104 AD) it was again re-established, in the fourth month of the fifth year it was stopped again and in the eleventh month re-established. Later the Emperors decreed the re-establishment of mathematics in the third year of the Dà Guān (大觀) reign (1109 AD) and in the third year of the Zhèng Hé (政和) reign (1113 AD). However, each time it was discontinued not long after. All this might be due to the 'general feeling that it was extravagant and did not really help in the running of the country'. In the time of the Southern Sòng Dynasty (1127–1280 AD), after the end of the Northern Sòng Dynasty, the subject of mathematics was discontinued once and for all and it was never re-established. Because of the on-and-off situation of mathematics in the Imperial Academy during the Northern Sòng period, very few famous mathematicians worth mentioning in history were produced. It was only education in the schools and the textbooks published by the Secretariat which made some contribution to the handing on of works of ancient mathematics.

One mathematician worth mentioning from the Northern Sòng period was Shěn Kuò (沈括 , 1031–1095 AD). Shěn Kuò's range of mathematics was very wide. He was well versed in mathematics and the sciences of the heavens. In his famous work *Dream Pool Essays* (夢溪筆談 , *Méng qi bi tǎn*) there were several passages concerning mathematical problems. Also among the State Astronomers at that time, besides Shěn Kuò, there were Chǔ Yǎn (楚衍), Zhū Jí (朱吉) and several others. They, too, were very good at calculations. Chǔ Yǎn's student Jiǎ Xiàn (賈憲) made outstanding con-tributions to the solution of equations.

In the year 1127 AD the Jīn (金) army captured the Northern Sòng capital Biàn Liáng (汴梁) (present-day Kāifēng, 開封), and the block print texts in the Secretariat were looted and destroyed. The various mathematical books were much affected by this. After the Jīn the Mongols rose up north of the Yangtze river and the Empire was split into North and South with the Southern Sòng in the South and the Mongols (who became the Yuán Dynasty) in the North.

Fortunately, however, in this period of confrontation, a very impor-tant new page was started in the history of mathematics in ancient China. In the South there was Qín Jiǔsháo (秦九韶) and Yáng Huī (楊輝) and in the North there were Lǐ Zhì (李治) and Zhū Shìjié (朱世杰). The writings of the four masters Qín, Lǐ, Yáng and Zhū fully reflect the splendid achievements of Chinese mathematics of this period. Their main writings were:

Qín Jiǔsháo: *Mathematical Treatise in Nine Sections* (數書九章, *Shùshū jiǔzhāng*), 18 segments (1247 AD);

Lǐ Zhì: *Sea Mirror of Circle Measurements* (測圓海鏡 *Cèyuán hǎijìng*), 12 segments (1248 AD); *New Steps in Computation* (益古演段 *Yìgǔ yǎnduàn*), 3 segments (1259 AD);

Yáng Huī: *A Detailed Analysis of the Mathematical Methods in the Nine Chapters* (詳解九章算法 , *Xiángjiě jiǔzhāng suànfǎ*), 12 segments (1261 AD; the surviving edition is not complete); *Computing Methods for Daily Use* (日用算法 , *Rìyòng suàn fǎ*), two segments (1262 AD; the surviving edition is not complete); *Yáng Huī's Methods of Computation* (楊輝算法 , *Yáng Huī suànfǎ*) seven segments (1274–1275 AD);

Zhū Shìjié: *Introduction to Mathematical Studies* (算學啓蒙, Suànxué qìmèng), seven segments (1299 AD); *Precious Mirror of the Four Elements* (四元玉鑑 , Sìyuán yùjiàn), three segments (1303 AD).

Apart from the two incomplete works by Yáng Huī all the rest of the above-mentioned works are complete and have survived to the present day.

From the point of view of the scope of the discussion and the difficulty of the types of problem involved these works are unsurpassed by the mathematical writings of any other period in ancient China. In them there are recorded many intellectual achievements of the highest order. Briefly, Qín Jiǔsháo's work records the solution of higher degree equations and the method of solution of simultaneous congruences. In their work Lǐ Zhì and Zhū Shìjié discuss the 'method of the celestial element' and the 'method of four unknowns'. They treat problems on the elimination of one or more unknowns from a system of equations and higher degree simultaneous polynomial equations. Yáng Huī's work principally reflects the situation of the mathematics commonly used in trading in the populace. In addition to these four masters the outstanding scientist Guǒ Shǒujing (郭守敬) (1231–1316 AD) of the Yuán period used the method of higher order differences in the computation of the *Works and Days Calendar* (授時曆 , *Shòu shí lì*) (1280 AD). This, too, was an outstanding achievement of the mathematicians of the Sòng–Yuán period. In the time of the Yuán Dynasty, besides the works of Yáng Huī, the following books in the field of commercial mathematics have survived to the present:

Transparent Curtain, Short Notes (透帘細草 , *Tòu lián xi cǎo*);
A Detailed Discussion on the Method of Calculations (詳明算法 , *Xiáng ming suàn fǎ*);
and *Dīng Jù's Arithmetical Methods* (丁巨算法 , *Dīng Jù suàn fǎ*)

In fact, during the Sòng and Yuán period, the mathematics of China surpassed that of Europe. The solution of higher degree equations was nearly 800 years earlier than Horner's method and the method of elimination of several unknown in systems of higher degree equations was 500 years earlier than Bézout's in Europe, the method of solution of simultaneous congruences was earlier by more than 500 years and the higher order interpolation method was earlier by nearly 400 years. Chinese mathematicians of those days were ahead in many important areas of mathematics. The mathematics of the Sòng and Yuán period is the most splendid page in the history of Chinese mathematics and it is also a large and colourful page in the world history of mathematics in medieval times.

We next briefly introduce the lives and work of the four famous mathematicians of the Sòng and Yuán period: Qín, Lǐ, Yáng, and Zhū.

It appears that Qín Jiǔsháo's book has not been translated into any European language, but there is an excellent study of Qín and his book by Libbrecht (1973). There is also an informative entry in the *Dictionary of Scientific Biography* (Ho Peng-Yoke 1970a) under 'Ch'in Chiu-shao'.

Qin Jiǔsháo's literary name was Dàogǔ (道古). He claimed to be from Lú county (魯) but in fact he was born in Sìchuān (四川) *ca.* 1200 AD. It was said in those days that he was 'by nature extremely clever, and skilful in astronomy, the theory of music, mathematics, and architecture'. According to his own description in the *Mathematical Treatise in Nine Sections:* 'In early years living with my parents in the central capital [the capital of the Southern Sòng, present-day Hángzhōu] I had the opportunity to learn from the State astronomers and also tried to learn mathematics from a hermit scholar'. Later he followed his father to Sìchuān and he himself held the post of district commander in Sìchuān and various other minor government posts. During that time the Mongol army was attacking Sìchuān province. In his preface to the *Mathematical Treatise in Nine Sections* he says:

. . . [I was] not caring about surviving in the midst of stone missiles, arrows and with the frequent threat of death. After ten years of hard life my spirit was spent and my morale was low. I believe all things can be accounted for by Numbers (數 , shù) and so I have been involved in searching for truth, seeking discussions with the learned ones, enquiring into the unknown and now I have a rough idea . . . [I am] attempting to formulate it in terms of questions and answers so as to make it available.

The *Mathematical Treatise in Nine Sections* was written in a period when soldiers were everywhere and horses were running wild, while Qin was struggling through a long period of hardship. 'Numbers' refers to *Yì Jīng* (see Legge 1882/1966, p. 365). In that philosophy as in Pythagorean philosophy (see Aristotle 1908/1952, Metaphysics 987b) everything had its 'number'.

The whole book of the *Mathematical Treatise in Nine Sections* is divided into nine parts and each part has nine problems: a total of 81 problems. Among these are:

1. Indetermine analysis. It describes the 'Great extension method of finding one' — the solution of simultaneous linear congruences.

2. Heaven and the seasons. This concerns the computation of the calendar and the measurement of rainfall, snowfall, etc.

3. Field boundaries. Measuring the area of fields.

4. Surveying. Right-angled triangles and double difference problems.

5. Corvée labour. Problems on 'Fair taxes' and other taxation problems (see Libbrecht 1973, p. 441).

6. Money and foodstuffs. Transportation of foodstuffs and capacities of warehouses.

7. Architecture. Problems on architectural constructions.

8. Military matters. Problems on the arrangement of tents and the supply of army necessities.

9. Commercial matters. Problems on the calculation of interest and on trade.

The *Mathematical Treatise in Nine Sections* contains some very complicated problems. For example in Chapter 8 the problem 'Measuring a circular fort from a distance' requires solving a tenth degree equation, and in Chapter 9, 'Repairing a fort and adjusting taxes' has 180 answers.

One can get some idea of Qín Jiǔsháo's philosophy of mathematics, his views on the objects of mathematics, and the relation between applications and various aspects of reasoning from his own preface to the *Mathematical Treatise in Nine Sections*. His view was 'the dào (道) and the Numbers do not have two different sources' and 'among all things there is none unrelated to Number'. Here 'dào' is used as in the Chinese Classic 道德經 (Dào dé Jīng, also written Tao Te Ching, see Legge 1891/1966) meaning the Way or the essence of the cosmos (cf. Libbrecht 1973, p. 57). This sort of idea is accurate in the way that it recognizes an aspect of quantity in things. However, starting from this line of thought can also lead to number mysticism, that is to say, this line of thought also has a misleading aspect. On the one hand Qín maintained that mathematics can 'manage wordly affairs, analogizes all created objects'; on the other he also claimed that mathematics can 'reach the deities, smooth the path of fate' and 'the changes of human affairs do not take place for no reason, the behaviour of deities and ghosts cannot be concealed'. Also he regarded 'reaching the deities, smoothing the path of fate' as important and 'managing wordly affairs, analogizing to all created objects' as less important. This is extremely inconsistent, in particular that when he reached further into mathematics until he got beyond the stage of 'enquiring into the unknown, feeling [he] had a rough idea' he still could not but acknowledge that 'mortals like me still cannot quite see what is called reaching the deities, smoothing the path of fate. As for the less important matter [that he had a rough idea of, i.e. mathematics] I carefully try to put [that] in the form of questions and answers so as to make it available.' That is to say, the 81 problems in the *Mathematical Treatise in Nine Sections* contain some of what is 'less important'. In addition Qín Jiǔsháo still put problems on numerology right at the beginning of the book. He also put the discussion of simultaneous first order congruences together with the numbers of *Dà Yǎn* (大衍) numerology of the *Great Appendix* (see Legge 1882/1966, p. 365, Libbrecht 1973, p. 354, also Needham 1959, p. 119 and note j) in the *Yì Jīng* and called this the dà yǎn qiu yī shù (大衍求一術 — Dà Yǎn method for finding one; that is, solving systems of linear congruences). Because of the restricted attitudes of that particular period, Qín Jiǔsháo could not have resolved this sort of contradictory thought by himself.

Contemporaneous with Qín Jiǔshào there was an outstanding mathematician, Lǐ Zhì (李治), in Northern China. The book *Sea Mirror of Circle Measurements* (測圓海鏡 , *Cèyuán hàijing*), written by Lǐ Zhì, is only one year later than the *Mathematical Treatise in Nine Sections* of Qín Jiǔsháo.

Unfortunately Lǐ Zhì's *Sea Mirror of Circle Measurements* has not been translated into a European language. However his book *New Steps in Computation* has been translated into French by van Hée (1913).

A detailed study of the *New Steps in Computation* has recently been published by Lam Lay-Yong and Ang Tian-Se (1984). See also the entry in the *Dictionary of Scientific Biography* (Ho Peng-Yoke 1970c) under Li Chih.

Li Zhi later changed his name, because he discovered it was the name of the third Táng emperor, to Lǐ Yě (李冶) by just removing one stroke in the second character. His literary name was Jìngzhāi (敬齋). He came from Luánchèng (欒城) in Zhēndìng (眞定), the present-day district around Shíjiāzhuāng (石家莊) in Héběi (河北) province (north of the Yellow river), was born in 1192 AD and was a magistrate in the city of Jūnzhōu (鈞州, present day Yù (禹)county) in Hénán (河南) province under the Jin Dynasty (1115–1234 AD). In 1232 AD Jūnzhōu was besieged and captured by the Mongol army. He fled to the north, hid in Tóngchuān (桐川), Tàiyuán (太原) and Píngdìng (平定) in Shānxī province (山西) and also went back to near his birthplace in Héběi province. He finally settled down in Yuánshì (元氏), at the foot of Mount Fēnglóng (封龍). Lǐ Yě was a famous scholar in the North in those days. Kublai Khan, the first Emperor of the Yuán dynasty, (1215–1294 AD, reigned 1260–1294 AD, who was also visited by Marco Polo), gave several audiences to him and several times he was given official posts, but he always resigned the positions soon after.

At Mount Fēnglóng, Lǐ Yě gave lectures in barns to a lot of people and some of those studied mathematics. Lǐ Yě died in 1279 AD at the age of 88 in Héběi province.

Lǐ Yě's work *Sea Mirror of Circle Measurements* has a total of 12 chapters, 170 problems. The problems all concern finding the diameters of circles inscribed in or cutting the sides of right-angled triangles from the lengths of the sides. Of all the mathematical works surviving to the present day the *Sea Mirror of Circle Measurements* is the first systematic work with a discussion of the 'celestial element' (天元, tiān yuán) method. Lǐ Yě's other work, called *New Steps in Computation* is based on a book *Ancient Works* (益古集, *Yigǔ ji*). In Lǐ Yě's own words: 'Rearranging and correcting passages, carefully redrawing diagrams, making hundreds who knew only a little now be able to satisfy themselves, how can I not be happy?' This is a revision of a work for beginners in the 'celestial element' method. The book has a total of three chapters, 64 problems.

It is also possible to understand Lǐ Yě's view of mathematics from the preface to the *Sea Mirror of Circle Measurements*. Lǐ Yě wrote:

Numbers are uncountable. I once tried to exhaust them by force but numbers do not obey orders, I could not get to the end and my strength was exhausted. Is it then true that numbers cannot be exhausted? Why cannot those numbers which have been given names be exhausted? So to say it is difficult to exhaust numbers, that is possible: to say that numbers cannot be exhausted, that is inconceivable. Again, in this world there is something that exists beyond existence. Beyond existence are the natural numbers. It is not only the unnatural numbers, it is also the reason for nature. So I once wanted to exhaust them by force but even with the resurrection of Lǐ Shǒu (隶首) [see Chapter 1, p. 1] it is still impossible to say what will happen. One can work with natural reasoning in order to understand the natural numbers, but then the mystery of the universe and of all other objects cannot be matched.

Here Lǐ Yě correctly points out that 'number' reflects objective existence. In a lot of interesting phenomena there is the 'existence of something beyond'; this is the 'natural numbers' and this is exactly the reflection of 'natural law'. They are 'exhaustible', not 'inexhaustible'; they can be known, it is not impossible for them to be known. At the same time because they are 'natural laws' it is only possible to work from their original form and we cannot 'exhaust [them] by force'. His point of view was very accurate. In the preface to the *New Steps in Computation* Lǐ Yě also criticized the view that derided mathematics as the 'dying branch of the Nine-nines'.

A short time after Qín Jiǔsháo and Lǐ Yě, there appeared the famous work of Yáng Huī in the Southern Sòng Dynasty (1127–1280 AD).

Yáng Huī (楊輝), otherwise known as Qiānguāng (謙光), was a native of Qiántáng (錢塘 , present-day Hángzhōu). The surviving material concerning his life is extremely limited. However, his written work is extremely important for understanding the development of mathematics at that time. Yáng Huī's work collected together the problems and methods of computation of several kinds of mathematical works that had long been lost. Some of the important methods of calculation, for example the earlier 'method of extracting roots by iterated multiplication' and 'the source of the method of extracting roots' (details below), were handed down to the present through Yáng Huī's work.

Lam Lay Yong (1977) has given a translation and beautiful commentary on *Yáng Huī's Methods of Computation*. She has also treated his *Computing Methods for Daily Use* (Lam Lay Yong 1972).

Improved methods of calculation by means of counting rods and some simplified calculation methods for multiplication and division are also recorded in Yáng Huī's work. These will be discussed in detail in Chapter 6. The 'Outline of the practice of calculation' presented in Yáng Huī's chapter *Alpha and Omega of Variations on Methods of Computation* (算法通變本末 , *Suàn fǎ tōng biàn běn mò*) in his book *Precious Reckoner for Variations of Multiplication and Division* (乘除通變算寶 , *Chéng chú tōng biàn suàn bǎo*) was a very popular and practical syllabus for learning mathematics at that time. This is very precious material for understanding mathematical education in general in the society of those days.

Just as for Yáng Huī, the material concerning the life of Zhū Shìjié (朱世杰) is very scanty. For the moment we can get a rough idea from the prefaces written by Mò Ruò (莫若) and Zǔ Yí (祖頤) to Zhū Shìjié's *Precious Mirror of the Four Elements* (四元玉鑑 , *Siyuán yùjiàn*).

Zhū Shìjié, whose literary name was Hànqīng (漢卿), was also known as Sōngtíng (松庭). Several times in his work it mentions 'Zhū Sōngtíng of

Mount Yàn' (燕山朱松庭) or 'Zhū Shìjié resident of Yàn', so his birthplace was very probably near present-day Běijīng. In the preface by Zǔ Yí it says: 'Mr Zhū Sōngtíng of Mount Yàn was a famous mathematician and he travelled widely throughout the country for more than 20 years. The people who came from many places to learn from him increased daily'. In Zǔ Yí's preface it also says: Zhū Shìjié 'travelled in all directions, he returned to Guǎnglíng (廣陵 — present-day Yángzhōu) and increasingly people gathered to learn from him'. From the available material it appears that among the mathematicians of ancient China Zhū Shìjié was the first professional mathematician who 'travelled in all directions'; he was also a professional educator in mathematics.

Lam Lay-Yong (1979) has recently published an account of Zhū Shìjié's *Introduction to Mathematical Studies*. Jock Hoe (1977) has published the first volume of his definitive study of the *Precious Mirror of the Four Elements* in French (van Hée 1932 is an incomplete translation into French of the *Precious Mirror of the Four Elements* and has a number of important omissions).

The three chapters of the *Introduction to Mathematical Studies* (*Suàn xué qí méng*, 算學啓蒙) of Zhū Shìjié, divided into 20 sections and 259 problems altogether, survive to the present. This book starts from computation methods for multiplication and division and goes on to root extraction and the method of the 'celestial element' (i.e. the solution of polynomial equations). It includes almost all of the various aspects and content of mathematics as a branch of science of that time. This book, a complete work going from the simple to the difficult, is indeed a very good textbook as an 'Introduction'. His other written work that survives to the present is the *Precious Mirror of the Four Elements*. The book as a whole is divided into three chapters, 24 sections, 288 problems. The main concern of the *Precious Mirror of the Four Elements* is the solution of systems of quadratic and higher degree equations. In the book there are seven problems on systems of equations in four unknowns, 13 problems on systems of equations in three unknowns, and 36 problems on systems of equations in two unknowns. The other important concern of the *Precious Mirror of the Four Elements* of Zhū Shìjié is the problem of finding the sums of series with a finite number of terms.

In the *Sequel to the Biographies of Mathematicians and Astronomers* (疇人傳續編 , *Chóu rén zhuàn xù biān*) of the Qīng Dynasty we find the following comments on Zhū Shìjié:

'Hànqīng of the Sòng and Yuán period, together with Qín Jiǔsháo and Lǐ Zhì (or Lǐ Yě) can be said to form a tripod. Qín contributed positive and negative and the extraction of roots, Lǐ contributed the celestial element and all those contributions stretch back into the past and will survive to thousands of future generations; Zhū includes everything, he has improved everything to such an extent that it is only understood by the gods and has surpassed the two schools of

Qín and Lǐ.' This sort of comment is in fact very appropriate. Workers in history of science in the West also maintain that Zhū Shíjié was 'one of the greatest mathematicians of his race, of his time, and indeed of all times . . . His *Precious Mirror of the Four Elements* is the most important book of its kind, and one of the outstanding mathematical books of mediaeval times.' (Sarton 1947, pp. 701, 703.)

5.2 The method of extracting roots by iterated multiplication — the method of solution of higher degree equations

1 The methods of extracting roots by iterated multiplication, the method of extracting cube roots

The achievement in the mathematics of the Sòng and Yuán period that should be mentioned first of all is the solution of higher degree equations.

It was mentioned in a previous chapter that complete methods of extracting square and cube roots already existed as early as in the *Nine Chapters on the Mathematical Art*. In fact, extracting square and cube roots is finding the solutions of $x^2 = A$ and $x^3 = B$. In ancient China the method of solution of ordinary equations was called the 'method of opening the square' (開方法, kāi fāng fǎ) because the methods of solution for ordinary equations all evolved from the method of extracting square roots. It was mentioned previously that the method of solving $x^2 + ax = b$ and other ordinary quadratic equations (which was called the 'corollary to extracting square roots') evolved from the method of extracting square roots; the method of solution $x^3 + ax^2 + bx = c$ and other ordinary cubic equations (which was called 'the corollary to taking cube roots') was evolved from the method of extracting cube roots. 'The corollary on extracting square roots' can be found in the *Nine Chapters on the Mathematical Art*, the *Mathematical Manual of Zhāng Qiujiàn* and 'the corollary on extracting cube roots' is recorded in the *Continuation of Ancient Mathematics*.

The mathematicians of the Sòng and Yuán period made a great advance on the 'method of opening the square' in ancient China, they solved the problem of extracting roots of arbitrary degree. Furthermore they solved the problem of finding solutions to polynomial equations of arbitrary degree.

Below we explain the developments in the period starting from Jiǎ Xiàn (賈憲).

Jiǎ Xiàn introduced a new type of method for extracting square roots and cube roots and on top of this the new method was generalized further to find roots of arbitrary degree.

Material on the life of Jiǎ Xiàn is very scarce. We only know that he was the student of Chǔ Yǎn (楚衍), an astronomer of the Northern Sòng period. The time when Jiǎ Xiàn did his research on mathematics is about the middle of the 11th century AD.

Jiǎ Xiàn's work has long been lost, but the methods are recorded in the

Table 5.1

shāng	商			a	$a + b$
shí	實	N	$N - a^3$	$N - (a + b)^3$
fāng	方			$3a^2$	$3(a + b)^2$
liàn	廉			$3a$	$3(a + b)$
yú	隅	1	1	1	
		(1)	(2)	(3)	

Reclassification of the Mathematical Methods in the 'Nine Chapters' (九章算法纂類 , *Jiǔzhāng suàn fǎ zuǎn lèi*) of Yáng Huì. These methods are called 'the method of extracting square roots by iterated multiplication' and 'the method of extracting cube roots by iterated multiplication', respectively. In order to introduce these new methods briefly we take the calculation of cube roots.

Suppose it is required to find $\sqrt[3]{N}$ and let the cube root be $a + b + c$ (for example $\sqrt[3]{N} = 234$, $a = 200$, $b = 30$, $c = 4$), then this problem is equivalent to solving

$$x^3 = N \qquad (x = a + b + c).$$

In the counting rod configuration for the old method in the *Nine Chapters on the Mathematical Art* going from top to bottom the five rows are the result (shāng, 商), the given number (shí, 實), the square (fāng, 方), the coefficient of x^2, literally, the side (lián, 廉), the coefficient of x^3, literally the borrowing rod (借算 , jiè suàn) (in the mathematics of the Sòng and Yuán period this last is no longer called the 'borrowing rod' but the 'corner' (yú, 隅) or 'bottom method' (下法, xià fǎ). To start with, only the given number and the corner are displayed as in (1) in Table 5.1. After getting the first place a of the root, knowing the coefficients in the expansion of $(a + b)^3$ are 1, 3, 3, 1, separately calculate a^3, $3a^2$, $3a$ and 1 and then continue as in (2) of Table 5.1 in order to find the second decimal place of the root. After finding the second place, b, of the root it is still necessary to work with the coefficients 1, 3, 3, 1 as above to calculate, in turn, $(a + b)^3$, $3(a + b)^2$, $3(a + b)$, 1 as in (3) of Table 5.1. Finally find the third place, c, of the root.

Jiǎ Xiàn's method of calculation is to take the procedure of calculating a^3, $3a^2$, $3a$, etc. and change it to first multiplying and then adding after each multiplication, so as to make all the calculations take place in one counting rod arrangement. We take the solution of $x^3 = N$ as $x = a + b + c$ as an example again: After getting the first place, a, of the root, Jiǎ Xiàn's method is 'using the result (a) multiply by the bottom row (1) getting the linear term (a); multiply the linear term by itself to form the square (that is, a multiplied by the linear term gives a square, namely a^2); take away the cube from the given number (that is, a is multiplied by the square and subtracted from the given number, giving $N - a^3$)': as in (1) and (2) in Table 5.2. (Note

Table 5.2

		a	a	a	a
shāng	商	a	a	a	a
shí	實	N	$N - a^2 \cdot a = N - a^3$	$N - a^3$	$N - a^3$
fāng	方	0	$0 + a \cdot a = a^2$	$a^2 + 2a \cdot a = 3a^2$	$3a^2$
liàn	廉	0	$0 + 1 \cdot a = a$	$a + 1 \cdot a = 2a$	$2a + 1 \cdot a = 3a$
yú	隅	1	1	1	1
(xià fǎ)	（下法）				
		(1)	(2)	(3)	(4)

(5)	(6)
$a + b$	$a + b$
$N - a^3$	$N - a^3 - [3a^2 + 3ab + b^2] = N - (a + b)^3$
$3a^2$	$3a^2 + (3a + b)b = 3a^2 + 3ab + b^2$
$3a$	$3a + 1 \cdot b = 3a + b$
1	1

(7)	(8)
$a + b$	$a + b$
$N - (a + b)^3$	$N - (a + b)^3$
$3a^2 + 3ab + b^2 + [3a + 2b]b = $ $= 3(a + b)^2$	$3(a + b)^2$
$3a + b + 1 \cdot b = 3a + 2b$	$3a + 2b + 1 \cdot b = 3(a + b)$
1	1

(9)	(10)
$a + b + c$	$a + b + c$
$N - (a + b)^3$	$N - (a + b)^3 - [3(a + b)^2 + 3(a + b)c + c^2]c = $ $N - (a + b + c)^3 = 0$
$3(a + b)^2$	$3(a + b)^2 + [3(a + b) + c]c$
$3(a + b)$	$3(a + b) + 1 \cdot c = 3(a + b) + c$
1	1

that what we have labelled as coefficients are *locations* (registers) on the counting board.) After this 'again take the root multiplied by the bottom row and enter it in the coefficient of x^2 (giving $2a$), multiply it [the root, a] by the linear term and add in to the coefficient of x row (giving $3a^2$)', as in (3) of Table 5.2. 'Again multiply by the bottom row and add it into the coefficient of x^2 (giving $3a$)', as in (4) and (5) of Table 5.2. To find the second place, b, similarly 'again take the second place of the root (that is b), multiply it by the bottom row and enter it into the coefficient of x^2 (giving $3a + b$); multiply the result (b) by the coefficient of x^2 and enter it into the coefficient of x row (giving $3a^2 + 3ab + b^2$); multiply by the result (b) and take it away from the given number [multiply by b and subtract from the given number, giving $N - (a + b)^3$]', as in (6) of Table 5.2; 'again using the second place of the root multiply the bottom row and enter into the coefficient of x^2 [giving $3a + 2b$] and multiply the result (b) by the coefficient of x^2 and enter it into the coefficient of x row [giving $3(a + b)^2$]', as in (7) of Table 5.2; 'again multiply the bottom row by (b) and add it into the coefficient of x^2 [giving $3(a + b)$]', as in (8) and (9) of Table 5.2. Finally 'using the third place (c) of the root, multiply it by the bottom row and enter it into the linear term [giving $3(a + b) + c$], multiply the coefficient of x^2 and add it into the coefficient of x^2 and add it into the coefficient of x [giving $3(a + b)^2 + [3(a + b) + c]c$]; multiply the coefficient of x and take away from the given number [giving $N - (a + b + c)^3$] and it turns out there is nothing left, so we have the number of the side of the cube (that is, the cube root)' as in (10) of Table 5.2.

One can clearly see that the difference between Jiǎ Xiàn's new method and the old method lies in replacing the squaring, cubing, and other procedures by adding after each multiplication. The method of extracting roots in ancient China used what is called the method of carrying and bringing over and in this way all the multiplications required in the process of adding after each multiplication use one and the same number (see also Lam Lay-Yong 1980 p 413, 1982).

The reason why this new method is so valuable is that it can easily be generalized for extracting roots of arbitrary degree (details below).

From the point of view of modern algebra the counting rod configurations corresponding to (1)–(5) in Table 5.2 are equivalent to proceeding to substitute $x = a + y$ after getting the first place a, and thereby altering the original equation $f(x) = 0$ to $\varphi(y) = 0$; configuration (5) simply presents the coefficients of the equation $\varphi(y) = 0$. In modern algebra this sort of substitution is commonly known as 'Horner's method'. The distinguishing feature of Horner's method is the addition after each multiplication. The calculations in this method are very similar to the 'method of combined division'. The calculation procedure for 'Horner's method' using $x^3 = N$ and $x = a + b + c$ as an example again is as below:

1	$+ 0$	$+ 0$	$- N$		
	a	a^2	$+ a^3$	$a + b + \ldots$	Corresponding
1	$+ a$	$+ a^2$	$-(N - a^3)$		to (1)–(5) in
	$+ a$	$+ 2a^2$			Table 5.2
1	$+ 2a$	$+ 3a^2$			
	$+ a$				
1	$+ 3a$	$+ 3a^2$	$-(N - a^3)$		
	$+ b$	$+ (3a + b)b$	$+ \{3a^2 + (3a + b)b\}b$		

$$1 + (3a + b) + \{3a^2 + (3a + b)b\} - [\underline{N - (a + b)^3}]$$

$$+ b \qquad\qquad + (3a + 2b)b$$

Corresponding
to (5)–(9) of
Table 5.2

$$1 + (3a + 2b) \qquad + \underline{3(a + b)^2}$$

$$+ \ b$$

$$1 + 3(\underline{a + b}) \qquad\quad + 3(a + b)^2 - [\underline{N - (a + b)^3}]$$

: :

It is clear that the calculation procedure in 'Horner's method' and the procedure in Jiǎ Xiàn's new method — the step-by-step solution in the 'method of extracting roots by iterated multiplication' — are exactly the same.

The English mathematician Horner published a paper (Horner 1819) introducing this method of calculation. The Italian mathematician Ruffini had earlier introduced a similar method (Ruffini 1804, Problem 3). In present-day textbooks this method is commonly called the 'Ruffini–Horner method'. But in fact as early as the middle of the eleventh century Jiǎ Xiàn wrote on this type of method and that is nearly 800 years earlier than in the West. Jiǎ Xiàn's 'method of extracting roots by iterated multiplication' is one of the most outstanding creations of ancient Chinese mathematics. The 'method of extracting roots by iterated multiplication' had a profound influence on the development of mathematics during the Sòng and Yuán period.

2 The diagram for 'the source of the method of extracting roots' — the table of binomial coefficients (Pascal's triangle)

Jiǎ Xiàn invented not only a new method for extracting square and cube roots, called 'the method for extracting roots by iterated multiplication' but also the method for extracting higher degree roots.

As in extracting square and cube roots one uses the formulae

$$(a + b)^2 = a^2 + 2ab + b^2$$
$$(a + b)^3 = a^3 + 3a^2b + 3ab^2 + b^3,$$

the method for higher degree roots (such as the fourth and fifth roots) uses the formulae

$$(a + b)^4 = a^4 + 4a^3b + 6a^2b^2 + 4ab^3 + b^4$$
$$(a + b)^5 = a^5 + 5a^4b + 10a^3b^2 + 10a^2b^3 + 5ab^4 + b^5$$

etc.

In these equations the most important aspect is knowing the coefficients of the terms in the expansion.

Jiǎ Xiàn not only gave the method for finding these coefficients. In Volume 16 344 of the *Great Encyclopaedia of the Yǒng Lè Reign Period*, which was looted during the Boxer Rebellion in 1900 and is now in

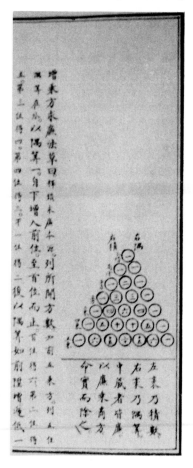

Fig. 5.1 The diagram for 'the source of the method of extracting roots' (Pascal's triangle).

Cambridge University Library, there is a diagram from Yáng Huī's book *A Detailed Analysis of the Mathematical Methods in the 'Nine Chapters'* (written in 1261 AD) entitled 'the source of the method of extracting roots', as in Fig. 5.1. Yáng Huī said that it 'appeared in the mathematical manual *The Key to Mathematics* (釋書, 鎖算 Sī shū suǒ suàn) Jiǎ Xiàn used this method'; that is to say, this diagram was invented by Jiǎ Xiàn in the eleventh century.

The diagram of the source of the method of extracting roots is an arrangement of numbers tabulated in a triangle, as follows:

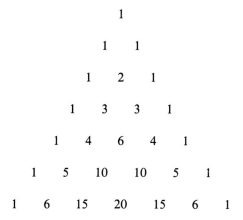

In it each row indicates the coefficients in the expansion of $(a + b)^n$. So the last row indicates the coefficients in the expansion of $(a + b)^6$:

$$(a + b)^6 = a^6 + 6a^5b + 15a^4b^2 + 20a^3b^3 + 15a^2b^4 + 6ab^5 + b^6.$$

A similar diagram can also be found in Zhū Shìjié's book *Precious Mirror of the Four Elements* (Fig. 5.2, p. 127). However, Zhū Shìjié had extended it to the eighth level. The title of the diagram in Zhū Shìjié's book refers to the 七 乘 方 (qī chéng fāng), literally the seven times multiplied: the first multiplying of *a* gives $a \times a = a^2$, the second gives $(a \times a) \times a = a^3$, etc. Therefore the seven times multiplied *a* is the eighth power of *a*. In Zhū Shìjié's diagram at the top left and right corners there are eight characters (中藏皆廉 , 開則橫視 , Zhòng cáng jiē lián, kāi zé héng shì) written vertically 'In the middle of the coefficients, [one] must read across'. 'In the middle of the coefficients' is to point out that the numbers in the middle part of the diagram are all the coefficients in the expansion of each binomial. 'Must read across' is to point out that each row indicates the appropriate coefficients to be used for a given exponent.

There are no problems in extracting higher degree roots using this 'source of the method of extracting roots'. So one can say: the appearance of this diagram is a signal that people had already grasped the method of extracting higher degree roots. In the West, historians of mathematics call this diagram, which is triangular in appearance, 'Pascal's triangle'. They claim that the French mathematician Pascal (1623–1662) was the first to invent it. This is not at all accurate. The Arab mathematician al-Kashi gave a table up to the ninth level in his *Key to Arithmetic* (1427). In China, after Zhū Shìjié, this 'source of the method of extracting roots' was also used by Wú Jìng (吳敬 , 1450 AD). In Europe it first appeared in printed form in 1527 AD on the front of a book by the German mathematician Apianus. Jiǎ Xiàn was almost 400 years earlier than al-Kashi and more than 500 years before Apianus.

In the *Great Encyclopaedia of the Yǒng Lè Reign Period* there is a passage following the diagram of the 'source of the method of extracting roots'. This passage explains the derivation of the numbers in each row in the diagram of the 'source of the method of extracting roots'. Below we present a translation of the original text. However the reader should note that this presents the formation of the triangle recursively. In order to understand the way the triangle is constructed each of the sentences below should be reworked from right to left, 1, 2, 3, 4, 5, 6. We shall explain this after giving the translation. Now we present the translation of the original. The original text, which can be seen in Fig. 5.1, says:

The method of finding the coefficients in iterated multiplication briefly says, the key to the source of finding the coefficients: DISPLAY THE DEGREE [literally root extraction number]. As before for the sixth degree, display five [middle] positions, the yú [陽] is [in] the outermost [position]. USING THE yú, 1, FROM BELOW ADD [it] IN TO THE PREVIOUS POSITION, CONTINUING TO THE FIRST POSITION, the first position getting 6, the second getting 5, the third getting 4, the fourth getting 3, and next getting 2. AGAIN USE THE yú AS BEFORE TO START, WORKING FROM ONE POSITION ABOVE.
    To get the second position:
6 old number; 5 add 10 then stop, 4 add 6 get 10, 3 add 3 get 6, 2 add 1 get 3.
    To get the third position:
6, 15 both old numbers, 10 add 10 then stop, 6 add 4 getting 10, 3 add 1 get 4.
    To find the fourth position:
6, 15, 20 all old numbers; 10 add 5 then stop, 4 add 1 get 5.
    To find the fifth position:
6 top coefficient; 15 second coefficient; 20 third coefficient; 15 fourth coefficient; 5 bottom coefficient add 1 getting 6.

The method used here is still that introduced by Jiǎ Xiàn in extracting square and cube roots — it is the method of adding after each multiplication — by 'iterated multiplication'. For example in finding the sixth root one uses the coefficients in the expansion of the sixth power of $(a + b)$. First of all display the fifth diagonal of the triangle where each entry is 1 as in (1) in Table 5.3. Then 'using the yú, 1, from below add it into the previous position continuing to the first position' as in (2) in Table 5.3. 'Again use the yú (1) as before to start, working from one position below', that is from bottom to top, each time using one coefficient less, as in (3) to (6) of Table 5.3. The final result, including the yú (1) and the leading coefficient (1) is just the coefficients in the expansion of $(a + b)^6$, namely 1, 6, 15, 20, 15, 6, 1. Obviously this 'method of extracting roots by iterated multiplication' can be used to obtain the coefficients in the expansion of $(a + b)^n$ for arbitrary degree n, so it is also possible to use these coefficients in order to extract roots of arbitrary degree.
    It is not difficult to deduce that this new method — 'the method of extracting roots by iterated multiplication' — can not only be used to work out the coefficients, but can also be applied directly to find roots of arbitrary degree. That is to say, one can use the actual *process* of calculating the triangle in

Table 5.3

	(1)	(2)	(3)	(4)	(5)	(6)	(7)
Bottom row	1	1 + 5 = 6 stop	6	6	6	6	6
2nd row	1	1 + 4 = 5	5 + 10 = 15 stop	15	15	15	15
3rd row	1	1 + 3 = 4	4 + 6 = 10	10 + 10 = 20 stop	20	20	20
4th row	1	1 + 2 = 3	3 + 3 = 6	6 + 4 = 10	10 + 5 = 15 stop	15	15
5th row	1	1 + 1 = 2	2 + 1 = 3	3 + 1 = 4	4 + 1 = 5	5 + 1 = 6 stop	6
yú (隅)	1	1	1	1	1	1	1

order to do the root extraction calculations without having to work out the triangle first. In fact it is just that. In *A Detailed Analysis of the Mathematical Methods in the Nine Chapters* by Yàng Huī contained in Volume 16 344 of the *Great Encyclopaedia of the Yŏng Lè Reign Period* there is a problem of obtaining a fourth root recorded: 'Given 1, 336, 336, what is its three times square root?'. Here 'three times square' means a number multiplied by itself again multiplied by itself and then again multiplied by itself, a total of three multiplications. Because there are three multiplications is called the three times square; in fact it is the fourth power. In Chinese mathematics books the nth power is always called the '$(n - 1)$ – times square' (cf. the remarks on Fig. 5.2 above). This problem of Yàng Huī's was also obtained from Jiǎ Xiàn's *Mathematical Manual*.

This problem is, in fact, finding $\sqrt[4]{1, 336, 336}$, and it is equivalent to solving the equation $x^4 = 1, 336, 336$, so $x = 34$.

The explanation in the book is as follows.

. . . getting the first number (place) of the root, multiply it by the [coefficient of x^4] xià fǎ (下法), getting the lower coefficient, multiply it by the lower coefficient getting the upper coefficient, multiply it by the upper coefficient getting the cube, multiply by the first number [of roots] and subtract it from the given number.

And

. . . using the first number multiply the xià fǎ and enter it in the lower coefficient, multiply the lower coefficient and enter it in the upper coefficient, multiply the upper coefficient and enter it in the square; again multiply the xià fǎ and enter it in the lower coefficient, multiply the lower coefficient and enter it in the upper coefficient; again multiply the xià fǎ and enter it in the lower coefficient. Square is 1, upper coefficient is 2, lower coefficient is 3, and xià fǎ is 4.'

And

. . . continue getting the root numbers [getting the second place of the root]. Multiply [the second place number by] the xià fǎ and enter it into the lowest coefficient, multiply the lower coefficient and enter it in the upper coefficient, multiply the upper coefficient and enter it into the cube, multiply by the [second place] number and subtract it from the given number, nothing is left. Thus obtaining one side of the three times square [that is, the fourth root]. (See also Lam Lay-Yong 1969, p. 85.)

In the above one multiplies the number below, and after the multiplication one adds it into the location above, each time stopping one position below the previous stopping place. This is using the 'method of extracting the root by iterated multiplication'. Obviously this type of method can not only be used for extracting roots of arbitrary degree but also for solving polynomial equations.

As early as the middle of the 11th century AD the problem of extracting

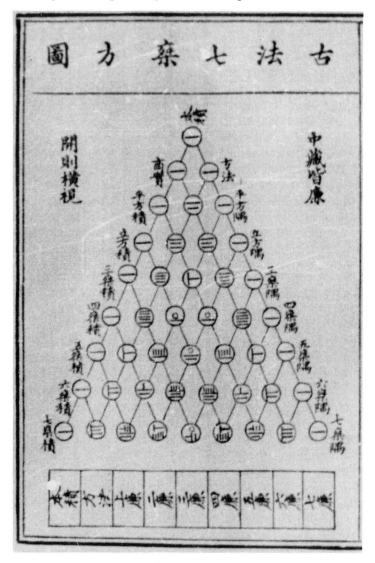

Fig. 5.2 The ancient method for extracting roots up to the eighth power.

roots of arbitrary degree had been solved. No-one can deny that this was an outstanding invention in the mathematics of ancient China. Not long afterwards the 'method of extracting roots by iterated multiplication' was generalized into a universal method for solving polynomial equations of arbitrary degree and within the three hundred years from the eleventh to the thirteenth centuries there were rapid developments. During this period the development of algebra marked a new peak in the mathematics of ancient China.

3   The solution of higher degree equations

Just as the 'corollary to extracting the square root' (for the solution of quadratic equations) and the 'corollary to extracting the cube root' (for the solution of cubic equations) evolved from the ordinary counting rod calculations for extracting square and cube roots, so it is not a difficult task to generalize the 'method for extracting square roots by iterated multiplication' and to extend it to the solution of higher degree equations in general. However, it should be noted that whether one is using the corollary to extracting the square root, the corollary to extracting the cube root or the method of extracting square roots by iterated multiplication, all of these require that the leading coefficient (that is, the coefficient of the highest degree term in the equation, which is the same as the yú or the xià fǎ in the counting rod arrangement for extracting square roots) be $+1$. The corollary on extracting square roots and the corollary on extracting cube roots also require the rest of the coefficients to be non-negative. So in order to be able to generalize the method of extracting roots by iterated multiplication for solving arbitrary higher degree equations it is necessary to break through this restriction.

According to the extant material Liu Yi (劉益) was the first to manage this. Liu Yi wrote a book entitled *Discussion on the Old Sources* (議古根源, *Yìgǔ gēnyuán*) which, however, is lost. But the *Practical Rules of Arithmetic for Surveying* (田畝比類乘除捷法 , *Tián mǔ bǐ léi chéng chù jiè fǎ*, 1275 AD) of Yáng Huī records more than twenty problems from Liu Yi's book *Discussion on the Old Sources*. In the preface of his book Yáng Huī wrote: 'Master Liu of Zhōngshān (中山) wrote a book *Discussion on the Old Sources*, . . . introducing the method of the corollary to extracting square roots independent of positive and negative [i.e. it does not matter whether the coefficients are positive or negative], which had never been heard of before'. Yáng Huī indicated that this was actually invented by Liu Yi. We know that Liu Yi worked around the second half of the twelfth or first half of the thirteenth century since Yáng Huī wrote of 'Master Liu Yi of Zhōngshān' and it was in 1113 AD that the government of the Northern Sòng Dynasty reclassified Dìngzhōu (定州) as Zhōngshān.

Of the 20 problems recorded in this book of Yáng Huī one equation is of fourth degree, four problems do not require extracting roots but just use multiplication and division and all the rest of the problems are quadratic equations — that is, problems of 'extracting square roots' and the 'corollary to extracting square roots'. These problems include the following types:

$$7x^2 = 9072 \ (x = 36)$$
$$x^2 - 12x = 864 \ (x = 36)$$

Table 5.4

			(1)	(2)	(3)	(4)	(5)
Root	商	Shāng	4	4	4	4	4
Constant term	實	Shí	4906	4906	4906	4906	$4906 - 1024 \cdot 4 = 0$
Linear term	三乘方法	Sān chéng fāng fǎ				$256 \cdot 4 = 1024$	1024
Upper coefficient (of x^2)	上廉	Shàng lián	128	128	$128 + 32 \cdot 4 = 256$	256	256
Lower coefficient (of x^3)	下廉	Xià lián	52	$52 + (-5) \cdot 4 = 32$	32	32	32
Yu (coefficient of x^4)	隅	Yú	−5	−5	−5	−5	−5

$$-x^2 + 60x = 864 \ (x = 24)$$
$$-5x^2 + 228x = 2595 \ (x = 24)$$
$$-3x^2 + 228x = 4320 \ (x = 36).$$

From the above one can see that the coefficients of various terms (including the coefficient of the leading term) are arbitrarily positive or negative. As far as the solution of quadratic equations in general is concerned Liu Yi had not yet used 'extracting roots by iterated multiplication' but had used two other methods: one of them is called 'extracting roots by accumulating products', the other is called the 'corollary to extracting square roots by subtraction'. These two methods are a little different from the method of extracting square roots in the *Nine Chapters on the Mathematical Art*. Whenever the coefficient is negative then the methods of calculating with negative numbers for addition, subtraction and multiplication are used.

The problem involving a fourth degree equation recorded in Yáng Huì's book is worthy of note. This is problem number 60 of Yáng Huì's book and it uses problem number 18 of Liu Yi's *Discussion on the Old Sources*. (See also Lam Lay-Yong 1977, pp. 130 and 269.) According to the counting rod configuration in the book, using modern mathematical notation, this problem is equivalent to solving the equation:

$$-5x^4 + 52x^3 + 128x^2 = 4096.$$

According to the original text this equation is solved as follows:

From the given numbers [by inspection] the first position of the root is 4 [as in (1) in Table 5.4], multiply by the yú, minus five, and subtract from the lower coefficient [of x^3] getting 32 [as in (2) in Table 5.4]. Using the first position of the root, four, of the fourth degree equation multiply the lower coefficient [i.e. of x^3] and add into the upper coefficient, total 256 [as in (3) in Table 5.4]. Again using the first position of the root, four, multiply the upper coefficient [now 256], getting 1024 as the coefficient of the linear term in the fourth degree equation [as in (4) in Table 5.4], using the first place of the root multiply the coefficient of the linear term, then subtract from the constant term, nothing left, giving the root, four [as in (5) in Table 5.4].

This is exactly the procedure of adding after each multiplication in the method of extracting roots by iterated multiplication.

The configuration of counting rods in the table above is equivalent to the Horner diagram

$$
\begin{array}{r}
-5 + 52 + 128 + 0 - 4096 \\
{} - 20 + 128 + 1024 + 4096 \\
\hline
-5 + 32 + 256 + 1024 0
\end{array} \quad \underline{4}
$$

but because the answer is just a single digit number the advantage of the method of extracting roots by iterated multiplication is not clearly shown.

However, one of the problems in Liú Yì's *Discussion on the Old Sources* introduced by Yáng Huī can be regarded as the earliest example of the use of the method of extracting roots by iterated multiplication for solving a higher degree equation.

The method of extracting roots by iterated multiplication developed slowly from the research of Jiǎ Xiàn and Liú Yì. By the middle of the thirteenth century AD the method had already been generalized into a method of solving equations of arbitrary degree in Qín Jiǔsháo's *Mathematical Treatise in Nine Sections* (1247 AD).

There are altogether more than 20 problems requiring the solution of equations in the *Mathematical Treatise in Nine Sections*. Among these are 20 problems on quadratic equations (including x^2 = A-type equations with just two terms), one cubic equation, four quartic equations and one tenth degree equation. They include the following examples:

$$4.608x^3 - 3000000000 \times 30 \times 800 = 0$$
($x = 25000$) (Problem 6-6, i.e. part 6 — problem 6 in Qín's book)
$$-x^4 + 763200x^2 - 40642560000 = 0$$
($x = 840$) (Problem 3-1)
$$-x^4 + 15245x^2 - 6262506.25 = 0$$
$$\left(x = 20\frac{1298025}{2362256} \right) \text{ (Problem 3-8)}$$
$$-x^4 + 1534464x^2 - 526727577600 = 0$$
($x = 720$) (Problem 4-6)
$$x^{10} + 15x^8 + 72x^6 - 864x^4 - 11664x^2 - 34992 = 0$$
($x = 3$) (Problem 4-5).

Here the coefficients may be positive or negative, integral or decimal. In a word, the 'method of extracting roots' has been generalized, it is applicable for finding solutions of various types of equation.

For most of the problems Qín Jiǔsháo's original text includes diagrams of the counting rod configurations which were used to explain each step in the method of extracting roots by iterated multiplication. For example, for Problem 3-1, the first problem of the third part of Qín's book, there are 21 diagrams used to illustrate the procedure for the solution of the quartic

$$-x^4 + 763200x^2 - 40642560000 = 0 \ (x = 840).$$

This is the problem called 'Finding the area of a pointed field'. We now take this as an example, but for the sake of simplicity we combine these 21 configurations into eight diagrams and at the same time we include the headings from the original diagrams alongside our eight diagrams. (See also Libbrecht 1973, pp. 181–189, but in Libbrecht's Diagram 29 for *c* read *e'*.)

1. The layout of the equation on the counting board.

− 40642560000	實 shí	constant term
0	方 fāng	linear coefficient (of x)
+ 763200	上廉 shàng lián	upper coefficient (of x^2)
0	下廉 xià lián	lower coefficient (of x^3)
− 1	隅 yú	coefficient of x

2. 'The upper coefficient moves over one place and the yú moves over three places whereas the root moves over one place. Then again move the upper coefficient over one place, the yú over three places and the root over one place. Assume the first position of the root to be 800.

8	商 shāng	root
− 40642560000	shí	
0	fāng	
+ 763200	shàng lián	
0	xià lián	
− 1	yú	

3. 'Using the first place of the root, multiply the yú and add into the lower coefficient. Using the first place of the root multiply the lower coefficient, cancelling from the upper coefficient (i.e. positive and negative cancel from each other). Using the first position of the root multiply the upper coefficient and enter it into the linear coefficient. Using the first position of the root multiply the linear coefficient and subtract the product obtained from the constant term. In adding the negative given number to the positive product of the first position of the root and the linear coefficient change the sign of the given number and subtract it from the product of the root and the linear coefficient, the result is positive and this is called "changing bones" (換骨, huàn gǔ).

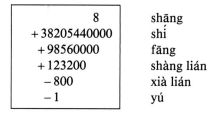

8	shāng
+ 38205440000	shí
+ 98560000	fāng
+ 123200	shàng lián
− 800	xià lián
− 1	yú

4. 'Using the first position of the root multiply the yú and enter in the lower coefficient: first change; using the first position of the root multiply the lower coefficient and enter in the upper coefficient, again cancelling. The opposite sign of the upper coefficient cancels the other. Using the root

multiply the upper coefficient and enter it into the linear coefficient and
cancel. The positive and negative entries in the linear coefficient cancel each
other.

8	shāng
+ 38205440000	shí
− 826880000	fāng
− 1156800	shàng lián
− 1600	xià lián
− 1	yú

5. 'Using the first position of the root multiply the yú and enter in the
lower coefficient: second change. Using the first position of the root multiply
the lower coefficient, enter in the upper coefficient.

8	shāng
+ 38205440000	shí
− 826880000	fāng
− 3076800	shàng lián
− 2400	xià lián
− 1	yú

6. 'Using the first position of the root multiply the yú, enter in the lower
coefficient: third change.

8	shāng
+ 38205440000	shí
− 826880000	fāng
− 3076800	shàng lián
− 3200	xià lián
− 1	yú

7. 'The linear term recedes one position, the upper coefficient two
positions, the lower coefficient three positions and the yú four positions; the
root stays fixed: fourth change.

8	shāng
+ 38205440000	shí
− 826880000	fāng
− 3076800	shàng lián
− 3200	xià lián
− 1	yú

8. 'Using the linear term divide the constant term, the second position is 40 [by inspection], multiply the yú and enter in the lower coefficient. Using the second place of the root multiply the lower coefficient and enter in the upper coefficient. Using the second place of the root, 40, multiply the linear coefficient and subtract it from the constant term, nothing is left. So we get the root to be 840 steps as the measure of the field.'

840	shāng
	shi
− 955136000	fāng
− 3206400	shàng lián
− 3240	xià lián
− 1	yú

It is evident that the special feature of this method of calculation is again the addition after each multiplication. In the explanation the procedure 'first change, second change, third change' is equivalent to making the substitution $x = 800 + y$ after obtaining the first position of the root, 800. This method of substitution by way of the process of addition after each multiplication is basically the same as Horner's method, which is used nowadays. For example in the present use of Horner's method the procedure involved in the substitution $x = 800 + y$ goes as follows:

$$
\begin{array}{l}
-\ 1\ +\ 0\ +\ 763200\ +\qquad\quad 0\ -\ 40642560000 \quad\big|\ 800 \\
\quad\ \ -\ 800\ -\ 640000\ +\ 98560000\ +\ 79048000000\quad\big| \\
\hline
-\ 1\ -\ 800\ +\ 123200\ +\ 98560000\ +\ 38205440000 \\
\quad\ \ -\ 800\ -\ 1280000\ -\ 925440000 \\
\hline
-\ 1\ -\ 1600\ -\ 1156800\ -\ 82688000 \ldots\ldots\ldots\ \text{(equivalent to 'first change')} \\
\quad\ \ -\ 800\ -\ 1920000 \\
\hline
-\ 1\ -\ 2400\ -\ 3076800 \ldots\ldots\ldots\ldots\ldots\ldots\ldots\ldots\ \text{('second change')} \\
\quad\ \ -\ 800 \\
\hline
-\ 1\ -\ 3200\ -\ 3076800\ -\ 826880000\ +\ 38205440000 \ldots.\ \text{('third change').}
\end{array}
$$

So after substitution the formula obtained is:

$$ -\ y^4\ -\ 3200y^3\ -\ 3076800y^2\ -\ 826880000y\ +\ 38205440000\ =\ 0. $$

Again, using the same form of calculation to solve it we get $y = 40$; adding this in $x = 800 + y$ we get $x = 840$.

The Horner's expression listed above is equivalent to the counting rod configurations in Steps 2, 3, 4, and 5, finally arriving at the configuration of Step 6.

In general the signs of the coefficients of the terms do not change upon the substitution $x = a + y$ but their absolute values should decrease. However, in particular cases, where there is a change of sign, Qín Jiǔsháo called this 'changing bones' and if the absolute value did not decrease but increased, it was called 'being born again' (投胎, tóu tāi). In the solution above of the problem of 'Finding the area of a pointed field' the situation of 'changing bones' occurs; that is, after getting the first position of the root, 800, signs of coefficients in the equation change from negative to positive in the process of substituting $x = 800 + y$. (See also the major article by Wang Ling and Needham 1955.)

5.3 From the 'technique of the celestial element' to the 'technique of four unknowns'

1 The origin and development of the 'technique of the celestial element' (天元術, tiān yuán shù)

In general the use of the technique of solving equations in order to find a solution to a practical problem can be divided into two steps. First of all one has to take an unknown number and then, according to the conditions given in the problem, find the equation that governs the unknown: this is step one. Then the second step is to solve the equation and find the unknown number. The mathematicians of the eleventh to the thirteenth centuries in ancient China not only invented the 'method of extracting roots by iterated multiplication', which is a general method for solving higher degree equations, they also invented the general method for obtaining equations from given conditions. This was given the name 'technique of the celestial element'.

Although Qín Jiǔshào's book *Mathematical Treatise in Nine Sections* has a systematic description of the 'method of extracting roots by iterated multiplication' it lacks a systematic method for the procedure for writing down the equation. Of the mathematical texts still in circulation, the first to give a systematic description of the 'technique of the celestial element' are the works *Sea Mirror of Circle Measurements* (1248 AD) and *New Steps in Computation* (1259 AD) written by Lǐ Yě. The 'technique of the celestial element' is also used to obtain equations in the works entitled *Introduction to Mathematical Studies* (1299 AD) and *Precious Mirror of the Four Elements* (1393 AD) by Zhū Shíjié. In particular, the 'technique of the celestial element' was generalized to the 'technique of four unknowns [four elements]' (四元術, sì yuán shù) in the book *Precious Mirror of the Four Elements*, that is, generalized from the solution of equations in one unknown to that of equations in four unknowns.

'Celestial element' means the unknown number in the problem, 'establish the celestial element as such and such' means 'let x be such and such'. In the 'technique of the celestial element' a polynomial or polynomial equation is usually indicated by using the character 元 (yuán, element) at the first degree term or the character 太 (tài) at the constant term.

Below we present Problem 2 of Chapter 7 of Lǐ Yě's *Sea Mirror of Circle Measurements* as an example in order to explain the general procedure in setting up equations in the 'technique of the celestial element'. The original text of the problem read: '[Assume there is a circular fort of unknown diameter and circumference,] person A walks out of the south gate 135 steps and person B walks out of the east gate 16 steps and then they see one another. [What is the diameter?]' Lǐ Yě gave five different solutions for this problem. Below we give the second solution putting the original text (in translation) on the left and using modern mathematical notation on the right to explain the procedures involved.

'Briefly put: Let one celestial element be the radius of the fort, lay it down and first add to it the southward steps getting the gǔ (股).

[Explanation.] Let x be the radius of the circular fort then side OA = $x + 135$, side OB = $x + 16$.

'Secondly put down the easterly steps getting the gōu (勾).

Finding the diameter of a circular fort.

'Multiply the gōu and gǔ together, getting

OA × OB = $(x + 135)(x + 16)$
= $x^2 + 151x + 2160$,

'Divide by the celestial element, getting the hypotenuse

Divide by x getting hypotenuse =
$x + 151 + 2160x^{-1}$
(∵ AB.OC = OA.OB = 2△ABO)

'Multiply this by itself, getting the square of the hypotenuse and place it on the left.

Multiply this by itself, getting
$$(\text{hyp})^2 = x^2 + 302x + 27121$$
$$+ 652320x^{-1}$$
$$+ 4665600x^{-2}$$
(left-hand side).

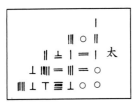

'Multiply the gōu by itself, getting

Again
$$OB^2 = (x + 16)^2$$
$$= x^2 + 32x + 256,$$

and again multiply the gǔ by itself, getting

$$OA^2 = (x + 135)^2$$
$$= x^2 + 270x + 18225.$$

'The two configurations added give

$$OB^2 + OA^2 = 2x^2 + 302x + 18481 = (\text{hyp})^2$$

which is the same value [as obtained before for the hypotenuse squared]. Cancel it with the [hypotenuse squared],

Equate to the left-hand side and simplify getting
$$-x^2 + 8640 + 652320x^{-1}$$
$$+ 4665600x^{-2} = 0,$$

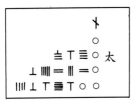

which is a quartic equation giving 120 steps as the radius of the fort.'

rationalizing the equation we get
$-x^4 + 8640x^2 + 652320x + 4665600 = 0.$

Solving, we get $x = 120$ (steps) as the radius of the circular fort.

On comparing the left- and right-hand sides of the above it is clear that, as a method of finding an equation, the 'technique of the celestial element' is roughly similar to the method used in present-day textbooks in algebra. In Lǐ Yě's book the derivation of the equation in the above example by cancellation after getting the two forms for the square on the hypotenuse is called 'cancelling the same number' or 'cancelling like results'.

In the above example the various configurations can be regarded as either equations or polynomials. From various points arising from the above example it is clear that the Chinese mathematicians of those days had already expertly grasped methods of calculation such as the addition, subtraction, multiplication and division of polynomials (for division it is confined to division by integral powers of x, that is, x, x^2, . . .). Also, the method of multiplying polynomials is carried out according to the 'method of extracting roots by iterated multiplication', that is, adding after each multiplication.

We must also point out that in those days all the equations involving the celestial element are expressed in the form of rational equations. In the case of irrational equations they always used squaring to eliminate the square root and rationalize; in the case of fractions they cross-multiplied and then solved the equation after making it linear. In the above example they did not use

$$\frac{(x + 135)(x + 16)}{x} = \sqrt{(2x^2 + 302x + 18481)} \ (= \text{hyp}),$$

but used

$$\left[\frac{(x + 135)(x + 16)}{x}\right]^2 = 2x^2 + 302x + 18481 \ (= \text{hyp}^2)$$

to get to 'cancelling like results'. Again, if the denominator of the left-hand side of the above equation was not x but $x^2 + bx + c$ or some other polynomial then the mathematicians of the Sòng and Yuán period always used cross-multiplication

$$[(x + 135)(x + 16)]^2 = (2x^2 + 302x + 18481)(x^2 + bx + c)$$

in order to proceed with the calculation. In conclusion, the final 'equation for extracting the root' derived by mathematicians of the Sòng and Yuán period was always in rational form. Because of limitations of space here we shall not give examples for each case.

From the above it is not difficult to see that the technique of the celestial element was already fully developed in the works of Lǐ Yě. So what about the development of the technique of the celestial element before the time of Lǐ Yě? Because there is very little material available we can only get a few clues from Zǔ Yǐ's preface to the *Precious Mirror of the Four Elements*. In his narrative on the very beginnings of the 'technique of the celestial element' Zǔ Yǐ says: 'Jiǎng Zhōu (蔣周) of Píngyáng district (平陽) [in present-day Shānxī (山西) province] wrote *Ancient Works* (益古, *Yì Gǔ*), Lǐ Wényī (李文一) of Bólù region (博陸) [in present-day Héběi], wrote *Raising Courage* (照胆, *Zhào Dǎn*), Shí Xìndào (石信道) of Lùquán district (鹿泉) [in present-day Héběi] wrote the *Seal Manual* (鈐經 *Qián Jīng*), Liú Rúxié (劉汝諧) of Píngshuǐ region (平水) [in present-day Shānxī province] wrote the *Full Explanation of Tabulated Equations* (如積釋鎖, *Rúji shisuǒ*) and Yuán Yù (元裕), a native of Jiàng (絳) [in present-day Shānxī province] wrote it up in detail and thus their successors knew about the celestial element.' Of the many books by these authors, Lǐ Yě quoted from the *Seal Manual* in his *Sea Mirror of Circle Measurements* and he also studied the book *Ancient Works* and then wrote his *New Steps in Computation*. So the mention of the stages of development of the technique of the celestial element in Zǔ Yǐ's preface is credible. From this we can put the date of the origin of the technique of the celestial element further back, to the beginning of the 13th century AD or even a little earlier. Indeed Needham (1959, p. 41) believes that the technique of the celestial element can be pushed 'well back into the + 12th century'.

One fact in particular should be noted: the development of the technique of the celestial element was quite local. Most of the works about the technique of the celestial element appeared in the present-day provinces of Héběi and Shānxī, around the east and the west parts of the Tàiháng (太行) Mountains. This region was the cultural and commercial centre in the time of the Jìn and Yuán dynasties (1115–1368 AD). The technique of the celestial element was born and developed there and naturally this is connected with the social and economic development at that time. So far no-one has found any similar development of the area of the technique of the celestial element in any other place.

Finally we must mention the way counting rods are used for the technique of the celestial element. For example, using the constant term N (太 , tài) as an indicator, the unknown can be above or below. Thus $ax^2 + bx + N = 0$ may be recorded as

either

a	
b	元 (yuán)
N	太

A form

or

N	太
b	元 (yuán)
a	

B form

For example Lǐ Yě's *Sea Mirror of Circle Measurements* uses the A form while *New Steps in Computation* uses the B form.

It is possible to use just the character yuán (元) or just the character tài (太) in recording. Thus $25x^2 + 280x - 6905 = 0$ can be recorded as:

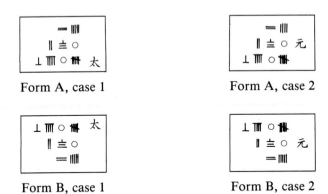

Form A, case 1   Form A, case 2

Form B, case 1   Form B, case 2

In general the mathematicians of the Sòng and Yuán periods used form B, case 2.

2   Zhū Shijié's 'technique of four unknowns'

The generalization of the 'technique of the celestial element' for equations with one unknown to systems of higher degree equations with two, three or four unknowns developed into the 'technique of four unknowns' (四元術, sì yuán shù). After the invention of the technique of the celestial element this was the most outstanding achievement of the Chinese mathematicians of the 13th century.

As mentioned above, the method of solving a system of linear equations in several unknowns is recorded in the *Nine Chapters on the Mathematical Art*. After people had grasped the technique of the celestial element it is natural that they extended it to systems of equations in several unknowns.

The developments from the technique of the celestial element to the technique of the four unknowns were mentioned briefly in Zǔ Yí's preface to the *Precious Mirror of the Four Elements*. After discussing the development of the 'technique of the celestial element' Zǔ Yí went on to say: 'Lǐ Dézài (李德載) of Píngyáng (平陽) region, having published *The Complete Collection for Heroes on Two Principles* (兩儀羣英集臻 , *Liǎng yì qún yǐng jizhēn*), they then had the earth unknown. The critic Liú Rùnfū (劉潤夫), the very talented student of Master Xíng Sōngbù (邢頌不) published *Heaven and Earth in a Bag* (乾坤括囊, *Qián kūn kuònáng*), which included two problems involving the human unknown (人元, rén yuán). My friend, Master Zhū Hànqīng [Zhū Shijié] of Yānshān district explained mathematics for many years. Having explored the mysteries of the three unknowns and sought out the hidden details of the *Nine Chapters*, he set up the four unknowns according to heaven, earth, man, and matter'. That is to say, to the celestial element [unknown] (x) (天元, tiān yuán) there were gradually added the earth element (y) (地元, dì yuán), the human element (z) (人元, rén yuán) and the material element (u) (物元, wù yuán).

Unfortunately the books of Lǐ Dézài and Liú Rùnfū mentioned in Zǔ Yí's preface have been long lost. Only Zhū Shìjié's *Precious Mirror of the Four Elements* survives to the present. This *Precious Mirror of the Four Elements* is a very valuable work for understanding the technique of the four unknowns.

There are brief accounts of the technique of four unknowns in the two prefaces to the *Precious Mirror of the Four Elements* written by Mò Ruò and Zǔ Yí. Mò Ruò's preface says:

. . . The method of the *Precious Mirror of the Four Elements* is to have the source of all the unknowns [the constant term] in the centre and to put the celestial [unknown] element at the bottom, the earth unknown on the left, the human unknown on the right and the material unknown at the top, opposites go up and down, proceed and recede to left and right, they interact and change criss-crossing with infinite variety. The hidden parts are revealed of excess and deficit, positive and negative rectangular tables, the method of taking powers and extracting roots; this is concise and deep. This information makes it possible to reveal the central ideas which were not worked out by our predecessors.

Then in Zǔ Yí's preface it says:

Set up four unknowns: heaven, earth, human and material. Use the source of the unknowns [the constant term] occupying the middle and set up the heaven in the gōu (勾), position [i.e. below] the earth in the gǔ (股) [on the left] the human on the hypotenuse [on the right] and the material at the zenith, look at the diagram and it is clear, ascending and descending, left and right proceeding and receding, criss-crossing positive and negative, dividing into four sorts. By way of varying, changing, excess counting rods interchanging positions and changing crossways, it is concise but not confused, is natural and naturally so, cancelling and then putting together so as to give the configuration for extracting the root.

These two passages give a brief description of the layout for the four unknowns configuration and the method of solution. Below we go a step further in explaining the presentation of the four unknowns using counting rods and the method of solution of systems of equations in four unknowns – essentially it is the method of elimination.

Regarding the display of the various coefficients for a celestial element configuration (see p. 139) as in a line from top to bottom, one can then say the technique of four unknowns displays the coefficients of various terms in the plane. Letting x, y, z, and u represent the four unknowns, the method of the four unknowns using counting rods is to put the constant term (太 tài) in the centre, the various coefficients of x below it, the various coefficients of y on the left, the various coefficients of z on the right, and the various coefficients of u above it. The coefficients of the products involving two unknowns (such as xy^2, z^3u^4, . . . etc.) are recorded at the corresponding points of intersection of two lines. The products of two non-adjacent unknowns as recorded in the corresponding holes of the grid as in the diagram below:

$....y^3u^2$	y^2u^2	yu^2	u^2	zu^2	z^2u^2	$z^3u^2....$
$....y^3u$	y^2u	yu	u	zu	z^2u	$z^3u....$
			$\boxed{yz}$			
$....y^3$	y^2	y		z	z^2	$\mathbf{z}^3....$
		$\boxed{xu}$				
$....xy^3$	xy^2	xy	x	xz	xz^2	$xz^3....$
$....x^2y^3$	x^2y^2	x^2y	x^2	x^2z	x^2z^2	$x^2z^3....$
$....x^3y^3$	x^3y^2	x^3y	x^3	x^3z	x^3z^2	$x^3z^3....$

So, for example, $x + y + z + u$ is recorded as

$$\boxed{\begin{array}{c} | \\ |\ 太\ | \\ | \end{array}}$$

and $(x + y + z + u)^2 = x^2 + y^2 + z^2 + u^2 + 2xy + 2xz + 2xu + 2yz + 2yu + 2zu$ is recorded as

$$\boxed{\begin{array}{ccccc} & & | & & \\ & \| & \circ_{\ \|} & \| & \\ |\ & \circ & _{\|}太^{\|} & \circ\ | & \\ & \| & \circ & \| & \\ & & | & & \end{array}}$$

Addition and subtraction of polynomials in four unknowns requires matching constant term with constant term and the coefficients of other terms with corresponding coefficients and then adding and subtracting the corresponding coefficients. For multiplication, first, for multiplication by an integral power of an unknown (e.g. x^3, y^4, etc.) multiply each term in the polynomial with four unknowns by the integral power of this unknown. For multiplying a configuration for four unknowns by the terms in a row or column, multiplication is equivalent to multiplying by each term in the configuration for four unknowns by each element of the row or column and then adding together the configurations for four unknowns obtained from each multiplication. Now for multiplying together two configurations for four unknowns just multiply one configuration by each row (or column) of the other configuration and then add up all the products. Division is restricted to dividing by integral powers of unknowns.

This method of presenting counting rod configurations for four unknowns and the four basic operations on them is exactly what was described in the prefaces by Mò Ruò and Zǔ Yí, which said: 'Have the source of the unknowns in the centre and put the celestial unknown at the bottom, the earth unknown at the left, the human unknown on the right and the material unknown at the top' and 'ascending and descending, left and right, proceeding and receding, criss-crossing positive and negative.'

We know that the key to the solution of systems of higher degree equations is the successive elimination of the unknowns from the equations in several unknowns until finally we get to one equation in only one unknown. This is what is generally called the 'method of elimination'.

At the beginning of Zhū Shìjié's book *Precious Mirror of the Four Elements*, four examples are given that have one unknown, two unknowns, three unknowns and four unknowns, respectively. After the third of these four examples he only describes the method of elimination of four unknowns tersely and concisely and in the other problems in the book *Precious Mirror of the Four Elements* there is no description or discussion of the method of elimination. The mathematicians Shěn Qīnpéi (沈欽裴, 1829 AD) and Luó Shìlín (羅士琳, 1836 AD) of the Qīng Dynasty gave different explanations of the method of elimination of the four unknowns.

There is no printed edition surviving of the *Commentary on the Precious Mirror of the Four Elements* (四元玉鑑細草, *Sìyuán yùjiàn xìcǎo*) by Shěn Qīnpéi but there is one complete and one incomplete manuscript in the Běijīng library. There is also a *Commentary on the Precious Mirror of the Four Elements* (四元玉鑑細草, *Sìyuán yùjiàn xìcǎo*) by Luó Shìlín of which the printed edition does survive and this has therefore had more influence. From another point of view this *Commentary* by Luó is not as good as that of Shěn in many respects.

Below we take the third of the above mentioned four problems 'Juggling three unknowns' as an example in order to illustrate the general ideas behind the 'method of elimination of four unknowns'. This is a problem with three unknowns. (See also Hoe 1977, pp. 172–211.) The system of equations to be solved is:

First formula

$$
\begin{array}{ccc}
-1 & 太 & -1 \\
 & 1 & \\
-1 & 0 & -1
\end{array}
$$

Second formula

$$
\begin{array}{ccc}
-1 & 太 & -1 \\
 & 1 & 1 \\
 & -1 &
\end{array}
$$

Three unknowns formula

$$
\begin{array}{ccccc}
1 & 0 & 太 & 0 & -1 \\
 & & 0 & & \\
 & & 1 & &
\end{array}
$$

This is equivalent to solving

$$
\begin{array}{ll}
-x - y - xy^2 - z + xyz = 0 & \text{First formula} \\
x - x^2 - y - z + xz = 0 & \text{Second formula} \\
x^2 + y^2 = z^2 & \text{Three unknowns formula.}
\end{array}
$$

Below we shall explain Zhū Shijié's methods in modern notation. The explanation given in Zhū Shijié's original text is: 'Using the second formula separate and eliminate. The human replacing the celestial unknown in the two formulae. The first formula gives [after substituting for y^2 from the third formula and dividing by x]

and the second gives

'Eliminate by reducing the hidden denominator (互隱通分 , hù yǐn tōng fēn), left gives

right gives

The inner columns multiplied by each other give

The outer columns give

Eliminate using the products of the inner and outer columns. Divide by four and obtain the formula

Solve the quartic equation getting hypotenuse [z] equals 5.'

In analysing the procedures of Zhū Shìjié's method of elimination of four unknowns, in general they can be classified as 'separating and eliminating', 'human replacing the celestial unknown', 'eliminating by reducing the denominators', 'inner two columns' and 'outer two columns'. Among these the inner two columns and the outer two columns are the final procedures in the method of elimination and they are also the most fundamental procedures in the method of elimination. These concern the solution of a system of equations with two unknowns in two columns (the simplest configuration for problems with two unknowns). In these lie the fundamental ideas behind Zhū Shijié's method of solution.

In the example of 'juggling three unknowns' the final elimination gives

'left

7	-6
3	-7
-1	-3
1	

, right

13	-14
11	-13
5	-15
-2	-5
	2

,

which is equivalent to having to solve

$$\begin{cases} (7 + 3z - z^2)x + (-6 - 7z - 3z^2 + z^3) = 0; \text{ left formula} \\ (13 + 11z + 5z^2 - 2z^3)x + (-14 - 13z - 15z^2 - 5z^3 + 2z^4) = 0; \\ \text{right formula.} \end{cases}$$

When the left and right formulae are put side by side $-6 - 7z - 3z^2 + z^3$ and $13 + 11z + 5z^2 - 2z^3$ are the 'inner two columns', $7 + 3z - z^2$ and $-14 - 13z - 15z^2 - 5z^3 + 2z^4$ are the 'outer two columns'.

After multiplying together the two inner and outer columns, subtracting and simplifying we have

$$4(-5 + 6z + 4z^2 - 6z^3 + z^4) = 0.$$

Dividing by four we get a quartic equation and solving it we arrive at the answer $z = 5$.

Taking the two left and right formulae and writing them in the ordinary way we have:

$$A_1 x + A_0 = 0$$
$$B_1 x + B_0 = 0$$

(A_0, B_1 are the inner two columns, A_1, B_0 are the outer two columns; A_0, B_1, A_1, B_0 are polynomials in z but do not involve x.) Thus multiplying the inner two columns and the outer two columns and subtracting leads to eliminating x, giving

$$F(z) = A_0 B_1 - A_1 B_0 = 0.$$

If the problem in two unknowns is not confined to two columns then it is necessary to use the method of 'reducing the hidden denominators'

(互隱通分, hù yǐn tōng fēn). For example:

$$A_2y^2 + A_1y + A_0 = 0 \tag{5.1}$$
$$B_2y^2 + B_1y + B_0 = 0 \tag{5.2}$$

(where A_2, A_1, A_0, B_2, B_1, B_0 are polynomials in x not containing y). Then multiplying eqn (5.1) by B_2 and eqn (5.2) by A_2 and subtracting we can eliminate the second powers of y, getting an equation of the form

$$C_1y + C_0 = 0. \tag{5.3}$$

Multiplying eqn (5.3) by y and using the same method with either eqn (5.1) or (5.2) we arrive at a form in two unknowns

$$D_1y + D_0 = 0. \tag{5.4}$$

Now using eqns (5.3) and (5.4) one can employ the procedure of 'inner two columns, outer two columns' and finally the result of the elimination is an equation with one unknown.

The problems on three or four unknowns first require 'separating and eliminating' in Zhū Shìjié's book. A lot of mathematicians explain the character 剔 (Tì) as 'Tì: separated into two', but they have different ways of looking at 'separated into two'.

Shěn Qīnpéi, a mathematician of the Qīng dynasty claims: 'taking the two configurations that require elimination, use the column (tài) 太 as marker'. Tì means divide into two, that is the column Tài belongs to the right-hand half. Then use the left half of eqn (5.1) after taking out the factor y and multiply by the right half of eqn (5.2); after taking out the factor y in the left-hand half of eqn (5.2) multiply the right half of eqn (5.1); subtract the two products and this permits the lowering of the degree of the y terms in the system of equations. Repeating this procedure we can eliminate all the terms in y. For example

$$A_2y^2 + A_1y + A_0 = 0 \tag{5.1}$$
$$B_2y^2 + B_1y + B_0 = 0 \tag{5.2}$$

(where A_2, A_1, . . ., B_0 contain x and z but not y)

Rewrite eqns (5.1) and (5.2) as

$$\begin{cases} (A_2y + A_1)y + A_0 = 0 & \text{(5.1a)} \\ (B_2y + B_1)y + B_0 = 0. & \text{(5.2a)} \end{cases}$$

Shěn Qīnpéi's method is equivalent to using A_0 from eqn (5.1) and B_0 from eqn (5.2) to multiply the polynomials in brackets in eqns (5.2a) and (5.1a), respectively, then after subtracting we get

$$C_1y + C_0 = 0 \tag{5.3}$$
(where $C_1 = A_2B_0 - A_0B_2$, $C_0 = A_1B_0 - A_0B_1$).

Using the same method again on eqns (5.3) with (5.1) or (5.2) gives an equation of the form

$$D_1 y + D_0 = 0. \tag{5.4}$$

Taking eqns (3) and (4) together (where C_1, C_0, D_1 and D_0 only contain x and z but not y) and using the same method to eliminate all the terms in y we eventually obtain a formula

$$E = 0$$

where E is a polynomial in x and z only.

This type of explanation by Shěn Qīnpéi is equivalent to regarding 'divided into two' as a further extension of 'reducing the hidden denominators'.

The other Qīng mathematician Luó Shìlín's explanation is equivalent to a sort of substitution method. We shall not describe it in detail here. When compared, Shěn Qīnpéi's way is somewhat more plausible.

Finally we return to explaining what is implicit in saying 'human replacing the celestial unknown'. In an equation with three unknowns, if the human unknown is eliminated it is not necessary to 'change the positions'. If the earth unknown is eliminated then the equation in the remaining two unknowns (celestial and human) must be changed using the 'tài' (太) marker as the centre so as to rotate the two displays of an equation in two unknowns from the fourth quadrant to the third quadrant. This is what is called 'human replacing the celestial unknown'.

In an equation with four unknowns, if the human unknown is eliminated then it is necessary to rotate the equation in the three unknowns (celestial, earth, and material) so as to make it occupy the third and fourth quadrants. This is what is called 'material replacing the celestial unknown'. In general the standard position for equations in two or three unknowns is in the third and fourth quadrants. If, for example, this is not the case then the method of 'replacing unknowns' is used so as to bring it into the standard form.

Obviously, 'replacing unknowns' is only moving the locations of the coefficients; the equation is not changed in substance.

The above is a brief description of Zhū Shíjié's method of elimination of four unknowns. In his time (13th century AD) the development of the techniques of the celestial element and of the four unknowns, which had started in northern China, reached its zenith. Because of the limitations of recording numbers using counting rods it is obvious that no more than four unknowns can be treated in this way. In Europe the method of elimination for higher degree equations was only described in a systematic way in the eighteenth century in the work of the French mathematician Bézout (1779 AD). This is five centuries after Zhū Shíjié (1303 AD).

5.4 The research into series by the mathematicians of the Sòng and Yuán period

1 The 'technique of small increments' by Shěn Kuò (沈括) and the problem of similar types of Yáng Hui

Very early on the mathematicians of ancient China had noted the problem of 'equal difference', i.e. arithmetic, series, and in the 11th to 13th centuries AD the mathematicians had gone a step further researching into higher order equal difference series and at that time they achieved outstanding results.

What is a higher order equal difference series? A series which, although the first differences of its terms may not be the same, yet has the differences of the differences the same everywhere is called a second order equal difference series. If the differences of the differences of the differences are the same it is called a third order equal difference series. For example for the series whose terms are 1, 4, 9, 16, 25, 36, 49, . . . the (first) differences $4 - 1 = 3$, $9 - 4 = 5$, 7, 9, 11, 13, . . . are not the same but the second differences $5 - 3 = 2$, $7 - 5 = 2$, $9 - 7 = 2$, . . . are all 2, so one can say that the series whose terms are 1, 4, 9, 16, . . . has constant second differences.

The earliest research into higher order equal difference series among the Sòng and Yuán mathematicians is from the Northern Sòng period starting with the 'technique of small increments' of Shěn Kuò (1032–1095 AD). The 'technique of small increments' is recorded in a section on 'techniques' by Shěn Kuò in the work *Dream Pool Essays* (夢溪筆談 , *Mèng qi bi tán*).

The 'technique of small increments' is commonly known as the problem of piling up stacks; as Shěn Kuò says 'piling qi (棋 , chess pieces), making stacks of jars or wine vessels in storehouses etc. Given a rectangular stack n layers high, with its top layer a objects wide and b objects long, bottom layer c objects wide and d objects long, what method should be used to find the total number of objects stacked up? Shěn Kuò's method of solution is 'Use the method of the 'rectangular platform' for the upper layers; for the bottom layers lay out the bottom width (c) and subtract it from the upper width (a), multiply the difference by the height (n), six into one, enter in the top row.' This is equivalent to the formula:

$$S = ab + (a + 1)(b + 1)$$
$$+ (a + 2)(b + 2) + \ldots + cd$$
$$= \frac{n}{6}[(2b + d)a + (2d + b)c] + \frac{n}{6}(c - a)$$

where in $\frac{n}{6}[(2b + d)a + (2d + b)c]$ the top rectangle's width and length are a and b, the bottom rectangle's width and length are c and d, respectively, and the height is n giving the volume of the 'rectangular platform' (cf. p. 44 above, (i)).

In fact Shěn Kuò's research opened up the way into higher order equal difference series for the next two or three hundred years. This research is all closely related to piling up stacks. This is just what was described by the Qīng mathematician Gù Guānguāng (顧觀光): 'The technique of stacking piles is detailed in Yáng (Huī)'s and Zhū (Shìjié)'s books but the invention should be credited to Shěn (Kuò).' [See Chapter 5 of Gù Guānguāng's 九數存古 (Jiǔ shù cún gǔ)].

There is a discussion of the problem of piling up stacks in the work of the Southern Sòng mathematician Yáng Huī. His details are a little more complete than Shěn Kuò's. There are four types of problem on piling stacks in his works *Detailed Analysis of the Mathematical Methods in the Nine Chapters* (1261 AD) and *Alpha and Omega of a Selection on the Applications of Arithmetical Methods* (1274 AD). He always put these at the end of the appropriate problems, as a way of introducing the problems on 'similar types. (cf. Lam Lay-Yong 1977, pp. 18 and 232.) For example:

1. 'Pile of fruit' is put at the end of 'rectangular pile' — this is the same as Shěn Kuò's 'rectangular platform':

$$S = ab + (a + 1)(b + 1) + \ldots + cd$$
$$= \frac{n}{b}[(2b + d)a + (2d + b)c] + \frac{n}{b}(c - a). \tag{5.5}$$

2. Also known as a 'pile of fruit', this problem is put after 'square-based pyramid' as a problem of similar type:

$$S = 1^2 + 2^2 + 3^2 + \ldots + n^2$$
$$= \frac{n}{3}(n + 1)(n + \frac{1}{2}). \tag{5.6}$$

3. 'Square pile' is put after 'square platform':

$$S = a^2 + (a + 1)^2 + (a + 2)^2 + \ldots + b^2$$
$$= \frac{n}{3}\left(a^2 + b^2 + ab + \frac{b - a}{2}\right). \tag{5.7}$$

4. 'Triangular pile' is put after 'tetrahedron':

$$S = 1 + 3 + 6 + 10 + \ldots + \frac{n(n + 1)}{2}$$
$$= \frac{n}{6}(n + 1)(n + 2) \text{ or}$$
$$= \frac{1}{6}n(n + 1)(n + 2). \tag{5.8}$$

Equation (5.5) is the same as Shěn Kuò's formula; eqn (5.6) is obtained by substituting $a = b = 1$, $c = d$ is eqn (5.5); eqn (5.7) is a particular case of eqn (5.5) when $a = b$ and $c = d$; and eqn (5.8) is a particular case of eqn (5.5) when $a = 1$, $b = 2$, $c = n$ and $d = n + 1$.

2 The 'technique of third order differences, linear, square, and cube' (平 , 立 , 定) (Píng, Lì, Dìng) in the *Works and Days Calendar* (授時曆 , *Shòu shí lì*) by Guō Shǒujìng (郭守敬) *et al.*

As well as the problems on piling up stacks, the research into higher order equal difference series has close connexions with problems on interpolation. It has already been mentioned in previous sections that in ancient China the problems on interpolation were closely connected with predicting the positions of the sun, moon, and the five known planets in computing the calendar.

If the observed motion of the sun is a quadratic function of time then it is easy to show that the path of the sun over equal time intervals has the form of a second order difference series. If the motion of the sun is a cubic function of time, then the path has the form of a third order difference series. Earlier we described how Liú Zhuó (劉焯) and Monk Yì Xíng (一行) independently gave formulae for second order interpolation for equal and unequal differences taking the motion of the sun as a quadratic. However this does not coincide with the facts: actually the motion of the sun is not a quadratic polynomial, it is a polynomial of higher degree. Yì Xíng also noticed this point but because of the inadequacies of the mathematics in those days he could not write the higher order difference interpolation formulae down accurately.

In the 13th century AD this problem was ingeniously solved by Wáng Xún (1235–1281 AD), Guō Shǒujìng (1231–1316 AD) and others. They produced the famous *Works and Days Calendar* (授時曆 , *Shòu shí lì* — 1280 AD), adopting the principle of third order interpolation to calculate the tables for the positions of the sun and moon. This is one of five major innovations in the *Works and Days Calendar*.

In the *Works and Days Calendar* the 88.91 days from the winter solstice to the spring equinox are divided into six equal periods each of 14.82 days. Let each period be ℓ. At the points $\ell, 2\ell, 3\ell, \ldots 6\ell$ the path of the sun is observed and then from each observation $\ell, 2\ell, 3\ell, \ldots 6\ell$ degrees are subtracted (the sun moves, on average, one degree each day and in $n\ell$ days it moves $n\ell$ degrees) to get the so-called 'accumulated differences', that is:

$$\text{Accumulated difference of } n\ell \text{ days} =$$
$$\text{actual degrees moved in } n\ell \text{ days} - n\ell.$$

Then using the differences of the accumulated differences successively we get the values of the 'first differences' (Δ). Next, successively subtracting the first differences (Δ) we get the 'second differences' (Δ^2). The values of the 'third differences (Δ^3)' can be found similarly; but the values of the 'fourth differences' are 0. Rewriting the table in the *Works and Days Calendar* in arabic numerals and at the same time exchanging rows and columns we get Table 5.5.

Table 5.5

	Accumulated days	Accumulated difference	First difference	Second difference	Third difference	Fourth difference
Initial period (0)	0	0				
			7058.0250			
First period (1)	14.82	7058.0250		– 1139.6580		
			5918.3670		– 61.3548	
Second period (2l)	29.64	12976.3920		– 1201.0128		0
			4717.3542		– 61.3548	
Third period (3l)	44.46	17693.7462		– 1262.3676		0
			3454.9866		– 61.3548	
Fourth period (4l)	59.28	21148.7328		– 1323.7224		0
			2131.2642		– 61.3548	
Fifth period (5l)	74.10	23279.9970		– 1385.0772		
			746.1870			
Sixth period (6l)	88.92	24026.1840				

N.B. One degree is taken as 10,000 divisions in the *Works and Days Calendar*. In the table 10,000.00 is one degree.

However Guō Shǒujìng and his co-workers did not make direct use of the formula for third differences according to the values listed above in making their calculations. They divided the accumulated differences by the number of days and called the result the 'average daily difference'. Next they composed a new table of average daily differences and then proceeded to calculate. Using average daily differences meant using only the formula for second order differences as shown by Table 5.5. In Table 5.5 for average daily differences, the third differences are 0. The reasoning involved is very simple. Let the accumulated difference at day x be $F(x)$, then, because the fourth differences in Table 5.5 are 0, $F(x)$ can be represented by a cubic $d + ax + bx^2 + cx^3$. But because the accumulated difference at the winter solstice is 0 (that is to say, $F(x) = 0$ at $x = 0$), we know that the constant term d in the cubic must be 0. This means that the average daily difference $F(x)$ can be written as a quadratic, namely $F(x) = a + bx + cx^2$. The various differences of the average daily differences can be listed as in Table 5.6.

In order to find the value of the average daily difference x days after the winter solstice, first take the x days and convert them to a fraction of a

Table 5.6

	Average daily difference[a]	First difference	Second difference	Third difference
First day after the winter solstice	[513.32]			
		[− 37.07]		
First period	476.25		[− 1.38]	
		− 38.45		0
Second period	437.80		− 1.38	
		− 39.83		0
Third period	397.97		− 1.38	
		− 41.21		0
Fourth period	356.76		− 1.38	
		− 42.59		0
Fifth period	314.17		− 1.38	
		− 43.97		
Sixth period	270.20			

[a] Average daily difference = $\dfrac{\text{accumulated difference}}{\text{number of days}}$

period, giving $\dfrac{x}{14.82}$. Then from the formula for the second differences from Table 5.5 one obtains:

Average daily difference x days after the

$$\text{winter solstice} = \frac{F(x)}{x}$$

$$= 513.32 + \frac{x}{14.82}\,(-37.07)$$

$$+ \frac{1}{2}\,\frac{x}{14.82}\left(\frac{x}{14.82} - 1\right)(-1.38).$$

Simplifying we get $\dfrac{F(x)}{x} = 513.32 - 2.46x - 0.0031x^2$.

Multiply by x days and one then gets the accumulated difference x days after the winter solstice = $F(x) = 513.32x - 2.46x^2 - 0.0031x^3$. That is to say the above mentioned $F(x) = ax + bx^2 + cx^3$, where a, b, c are as follows:

$$\begin{cases} a = 513.32 \\ b = -2.46 \\ c = -0.0031. \end{cases}$$

Putting $x = 1, 2, 3, 4, \ldots$ in turn and substituting in $F(x)$ should at once give the successive accumulated daily differences. When we substitute $1, 2, 3, 4, \ldots$ in turn it is at once clear that the as in $F(1), F(2), F(3), F(4) \ldots$ increase as multiples of $1, 2, 3, 4, \ldots$; the bs as $1^2, 2^2, 3^2, 4^2, \ldots$ and the cs as $1^3, 2^3, 3^3, 4^3, \ldots$. So Guō Shǒujìng *et al.* called a the 'linear difference' (定差, Dìng chā), b the 'square difference' (立差, Lì chā), and c the 'cube difference' (平差, Píng chā). Later on the mathematicians of the Qīng period also called this method of accumulated differences the 'technique of linear, square, and cube differences' (平立定三差術, Píng lì dìng sān chā shù).

In carrying out the calculations to find the accumulated differences for each day, Guō Shǒujìng and his co-workers did not make use of the formula substituting the number of days but still used the method of tabular calculations. From

$$F(0) = 0,$$
$$F(1) = a - b - c,$$
$$F(2) = 2a + 4b + 8c,$$
$$F(3) = 3a - 9b - 27c,$$
$$\cdots\cdots\cdots$$

It is not difficult to compute Table 5.7. From this table it is easy to obtain

$$\Delta^3 F(0) = -6c = -0.0186,$$
$$\Delta^2 F(0) = \Delta^2 F(1) - \Delta^3 F(0)$$
$$= -2b - 12c + 6c$$
$$= -2b - 6c$$
$$= -4.9386,$$
$$\Delta F(0) = \Delta F(1) - \Delta^2 F(0)$$
$$= F(1) - F(0)$$
$$= a - b - c$$
$$= 510.8560.$$

Table 5.7

Winter solstice day	$F(0) = 0$			
		$[\Delta F(0)]$		
Day 1	$F(1) = a - b - c$		$[\Delta^2 F(0)]$	
		$a - 3b - 7c$		$[\Delta^3 F(0)]$
Day 2	$F(2) = 2a - 4b - 8c$		$-2b - 12c$	
		$a - 5b - 19c$		$-6c$
Day 3	$F(3) = 3a - 9b - 27c$		$-2b - 18c$	
		$a - 7b - 37c$		$-6c$
Day 4	$F(4) = 4a - 16b - 64c$		$-2b - 24c$	
		$a - 9b - 61c$		
Day 5	$F(5) = 5a - 25b - 125c$			

After obtaining $F(0)$, $\Delta F(0)$, $\Delta^2 F(0)$, and $\Delta^3 F(0)$ it is easy to list the accumulated differences day by day. That is to say, knowing

	Accumulated difference	First difference	Second difference	Third difference
Winter solstice	$F(0) = 0$			
Day 1		$\Delta F(0) =$ 510.8560	$\Delta^2 F(0) =$ -4.9386	$\Delta^3 F(0) =$ -0.0186

Day 2

then by successively adding from right to left it is easy to reconstruct the lines in Table 5.8. This is the final table of accumulated differences obtained in the *Works and Days Calendar*, the so-called 'Works and Days Calendar system table' (授時曆立成 , Shòu shí lì lì chèng). The 'system table' is simply the table of numbers.

Summing up, Guō Shǒujing and his co-workers calculated the 'three differences: linear, square, and cube' (that is calculated a, b, and c) from the formula for second difference, then with these calculated the various orders of differences at the winter solstice finally listing the accumulated differences day by day. Although they did not write down the third order interpolation formulae, one can see from the calculations in the tables presented that they had firmly grasped the principles of third order interpolation and it is not difficult to generalize this into an interpolation method for higher degree polynomials.

Table 5.8

	Accumulated difference	First difference	Second difference	Third difference
Winter solstice	0			
		510.8560		
Day 1	510.8560		-4.9386	
		505.9174		-0.0186
Day 2	1016.7734		-4.9572	
		500.9602		-0.0186
Day 3	1517.7336		-4.9758	
		. . .		-0.0186
. . .	. . .		. . .	

3   Zhū Shìjié's work on stacking and finite differences

Among the research into stacking and finite differences by the mathematicians of the Sòng and Yuán periods, Zhū Shìjié's *Precious Mirror of the Four Elements* must be regarded as containing the most important results. In a certain sense Zhū Shìjié brought to completion the research work in this area by the Sòng and Yuán mathematicians. The results of this research by Zhū Shìjié are recorded in three sections of his book: 'Section on the arrangement of reeds' (total of seven problems), 'Finite differences' (five problems), and 'Stacking up fruit' (20 problems). (See also Hoe 1977, pp. 300 ff.)

Zhū Shìjié's *Precious Mirror of the Four Elements* mainly explains the technique of the celestial element and the technique of four unknowns, so at first glance there seems to be no system in the discussion on problems of stacking and of finite differences. However comparing all the problems and making a few inferences it is not difficult to discover that it has its own internal system.

As mentioned earlier, the problems on stacking are problems on finding the sums of higher order difference series. From Zhū Shìjié's many problems on finding the sums of series one draws the conclusion that they are a series of formulae with an important principle. Using present day mathematical notation to write them down this series of formulae is equivalent to:

$$1 + 2 + 3 + 4 + \ldots + n = \sum_{r=1}^{n} r = \frac{1}{2!} n(n + 1).$$

(Zhū Shìjié called this a 'pile of reeds' 茭草積 , jiāocǎo jī).

$$1 + 3 + 6 + 10 + \ldots + \frac{1}{2} n(n + 1)$$

$$= \sum_{r=1}^{n} \frac{1}{2} r(r + 1) = \frac{1}{3!} n(n + 1)(n + 2).$$

('Triangular pile' 三角垛 , sānjiǎo duǒ).

$$1 + 4 + 10 + 20 + \ldots + \frac{1}{3!} n(n + 1)(n + 2)$$

$$= \sum_{r=1}^{n} \frac{1}{3!} r(r + 1)(r + 2)$$

$$= \frac{1}{4!} n(n + 1)(n + 2)(n + 3).$$

(Zhū Shìjié called this a 'pile of scattered stars' 撒星形垛 , sǎxīng xíng duǒ or 'pile of superimposed triangles' 三角落一形垛 , sānjiǎo luòyīxíng duǒ. We explain the construction of the pile below.)

$$1 + 5 + 15 + 35 + \ldots + \frac{1}{4!} n(n + 1)(n + 2)(n + 3)$$

$$= \sum_{r=1}^{n} \frac{1}{4!} r(r + 1)(r + 2)(r + 3)$$

$$= \frac{1}{5!} n(n + 1)(n + 2)(n + 3)(n + 4).$$

(Zhū Shijié called this a 'triangular pile of scattered stars' 三角撒星形垛 , sānjiǎo sǎxing xíng duǒ or 'pile of superimposed scattered stars') 撒星更落一形垛, sǎxing gèng luòyixing duǒ.)

$$1 + 6 + 21 + 56 + \ldots$$

$$+ \frac{1}{5!} n(n + 1)(n + 2)(n + 3)(n + 4)$$

$$= \sum_{r=1}^{n} \frac{1}{5!} r(r + 1)(r + 2)(r + 3)(r + 4)$$

$$= \frac{1}{6!} n(n + 1)(n + 2)(n + 3)(n + 4)(n + 5).$$

(Zhu Shijie called this a 'triangular pile of superimposed stars' 三角撒星更落一形垛 , sānjiǎo sǎxing gèng luóyixing duo.)

By pile of 'superimposed' shapes Zhū Shijié means a stack built up as follows. For simplicity we consider a triangular pile. First consider triangles

T_1 T_2 T_3 T_4

Where the rth triangle has side r. Let layer L_1 be T_1. Now superimpose the first triangle T_1, in fact this is L_1, on the second triangle T_2 to get layer L_2. Next superimpose layer L_2 on triangle T_3 to get L_3. Continue in the same way to get $L_4, L_5, \ldots$.

L_1 L_2 L_3

The series of formulae can all be summarized in one formula:

$$\sum_{r=1}^{n} \frac{1}{p!} r(r + 1)(r + 2) \ldots (r + p - 1)$$

$$= \frac{1}{(p + 1)!} n(n + 1)(n + 2) \ldots (r + p - 1)(r + p).$$

Putting $p = 1, 2, 3, 4, \ldots$ one obtains the series of formulae one by one. For convenience we call these kinds of formulae 'triangular stacking' formulae.

It is worth noticing that the sum given by the preceding formulae is just the general term of the successive formulae in this series of triangular stacking formulae. From the above mentioned interpretation of stacking, the rth term in the subsequent formula (the general term) is just the sum of the previous r terms of the preceding formula, that is to say, take the total in the stack of the r layers indicated in the previous formula and put them all 'superimposed into one layer' (落爲一層 , luò wěi yì céng). This then becomes the rth layer of the stack for the next formula. Thus the second term of the formula with $p = 3$ is formed by taking the first two terms of the previous formula where $p = 2$ and 'superimposed into one layer'; the third term is formed from the first three terms of the formula with $p = 2$ 'superimposed into one layer'; this is perhaps the reason Zhū Shìjié described the succeeding formula as the previous formula 'superimposed into one layer'. Thus:

'Triangular pile' ($p = 2$) is called 'pile of reeds ($p = 1$) superimposed into one layer';

'pile of scattered stars' ($p = 3$) is also called 'pile of superimposed triangles';

'triangular pile of scattered stars' ($p = 4$) is also called 'pile of super-imposed stars'. Finally $p = 5$ gives a 'triangular pile of superimposed scattered stars'.

As well as the system for triangular stacking the series of formulae below can also be regarded as another system. Again they can be summarized by the following formula:

$$\sum r \, \frac{1}{p!} r(r + 1)\,(r + 2) \ldots (r + p - 1)$$

$$= \frac{1}{(p + 2)!} \, n(n + 1)\,(n + 2) \ldots (n + p)(\overline{p + 1}\cdot n + 1).$$

That is to say, multiply each term of the formula for triangular stacking by its corresponding number r and then solve the problem of finding the sum of the proposed series. When p takes the values 1, 2, 3, the above formula yields:

$p = 1$ $1.1 + 2.2 + 3.3 + \ldots n.n = \displaystyle\sum_{r=1}^{n} r.r = \frac{1}{3!}\, n(n + 1)\,(2n + 1)$

(Zhū Shìjié called this 'square pile' 四角垛 , Sìjiǎo duǒ.)

$p = 2$ $1.1 + 2.3 + 3.6 + \ldots + n.\dfrac{1}{2}\, n(n + 1)$

$\qquad\qquad = \displaystyle\sum_{r=1}^{n} r.\frac{1}{2!}\, r(r + 1)$

$$= \frac{1}{4!} \, n(n + 1) \, (n + 2) \, (3n + 1)$$

('Peaked pile' 嵐峯形垛 , lánfēng xíngduǒ.)

$p = 3$ $1.1 + 2.4 + 3.10 + \ldots + n.\frac{1}{3!} \, n(n + 1) \, (n + 2)$

$$= \sum_{r=1}^{n} \, r.\frac{1}{3!} \, r(r + 1) \, (r + 2)$$

$$= \frac{1}{5!} \, n(n + 1) \, (n + 2) \, (3n + 3) \, (4n + 1)$$

('triangular peaked pile' 三角嵐峯形垛 , sānjiǎo lánfēng xíngduǒ or 'peaked pile superimposed into one layer' 嵐峯更落一形垛 , lánfēng gèng luòyīxing duǒ). We can call these systems the systems of 'peaked piling'.

In addition to the two systems mentioned above, Zhū Shijié also has other sorts of problems on finding the sum of series, for example he takes the formula for triangular stacking, in turn multiplies each term by the corresponding term of an equal difference series and then solves the problem of finding the sum. We do not pursue this here because of limitations of space but see Hoe (1977, p. 307) for further details.

To fail to mention Zhū Shijié's achievements in the area of finite differences would be too great an omission. The treatment of finite differences is one of the most outstanding parts of Zhū's book *Precious Mirror of the Four Elements*. Because Zhū Shijié had already grasped the series of formulae for triangular piles he was in a position to go on further than his predecessors.

In the area of the 'calculus of finite differences' there are altogether five problems in the *Precious Mirror of the Four Elements* all connected with the method of finite differences. Here Zhū Shijié was the first person in the history of mathematics in China and even in the world history of mathematics to introduce an accurate formula for finite differences, including fourth order differences.

In Zhū Shijié's explanation of the last problem in the area of the 'calculus of finite differences' there is appended a problem posed by Zhū Shijié together with the method of solution that he gave. This question and answer implicitly describe the whole content of Zhū Shijié's method for finite differences. The problem given in the explanation is

. . . using the cubic [i.e. third order] method in recruiting soldiers, initially recruit 3 to the power 3 [the first day 3^3 are recruited.] then recruit increasing the side by one [the second day $(3 + 1)^3 = 64$ are recruited, the third day $(3 + 2)^3 = 125$ are recruited] . . . . Now 15 cubed are recruited and each day each man is paid 250 copper cash, what is the number of soldiers recruited and the amount of money paid out?

Table 5.9

	First difference Δ	Second difference Δ^2	Third difference Δ^3	Fourth difference Δ^4
$f(0) = 0$				
	$f(1) - f(0) = 3^3 = 27$			
$f(1) = 3^3$		37		
	$f(2) - f(1) = 4^3 = 64$		24	
$f(2) = 3^3 + 4^3$		61		6
	$f(3) - f(2) = 5^3 = 125$		30	
$f(3) = 3^3 + 4^3 + 5^3$		91		6
	$f(4) - f(3) = 6^3 = 216$		36	
$f(4) = 3^3 + 4^3 + 5^3 + 6^3$		127		$\ldots$
	$f(5) - f(4) = 7^3 = 343$		$\ldots$	
$f(5) = 3^3 + 4^3 + 5^3 + 6^3 + 7^3$		$\ldots$		

Let $f(n)$ indicate the total number of soldiers recruited in n days. Then from the data in the question one can construct the finite difference table shown in Table 5.9.

The explanation in Zhŭ Shijié's original text says:

Find the first difference is 27 [that is $\Delta f(0)$], second difference 37 [that is $\Delta^2 f(0)$], third difference 24 [$\Delta^3 f(0)$], last difference 6 [$\Delta^4 f(0)$]. To find the number of recruits: use the recruiting period to the present [that is the number of days, n, of recruiting soldiers] as the first coefficient; subtract one from the recruiting period and form a pile of reeds [for that number] $\left(\frac{1}{2!} n(n-1)\right)$ as the second coefficient; subtract two from the recruiting period and form a triangular pile $\left(\frac{1}{3!} n(n-1)(n-2)\right)$ as the third coefficient; subtract three from the recruiting period and form a pile of superimposed triangles $\left(\frac{1}{4!} n(n-1)(n-2)(n-3)]\right)$, as the last coefficient. Using the differences multiply each pile and add the four terms to get the number of soldiers recruited.

In modern algebraic notation Zhū Shijié's explanation is equivalent to the formula below:

$$f(n) = n\Delta + \frac{1}{2!} n(n-1)\Delta^2$$

$$+ \frac{1}{3!} n(n-1)(n-2)\Delta^3$$

$$+ \frac{1}{4!} n(n-1)(n-2)(n-3)\Delta^4.$$

This formula is exactly the same as the Newton formula in common use at present and is completely accurate.

Zhū Shijié clearly points out that each of the coefficients in the formula given above is just a triangular pile (that is, of the first type mentioned above). This is his outstanding contribution. Because he accurately pointed out this feature one can maintain that Zhū Shijié could have generalized his formula to finite difference formulae of arbitrarily high degree and can opine that he had already grasped the finite difference method for arbitrary degrees. From this it is possible to work out how to solve all the problems of finding the sums of higher order equal difference series using the method of finite differences.

Here Zhū Shijié surpassed the ancient Chinese mathematics and in particular the method of finite differences used by Wáng Xún (the Astronomer Royal) and Guō Shǒujing *et al.* in *Works and Days Calendar* in the Yuán dynasty, and ascended to an entirely new level. From one point of view one can say that Zhū Shijié had finally completed the research work in this area by the mathematicians of ancient China.

In Europe it was the British astronomer Gregory who first applied the method of finite differences but it was only in the work of Newton (1676–1678 AD) that the general formula for finite differences first appeared. In this area Zhū Shijié's work preceded that of his European counterparts by 400 years.

5.5 Achievements in other areas

1 'Indeterminate analysis'

'Indetermine analysis' means solving the problem of linear congruences. This sort of problem can be dated back into ancient Chinese mathematics. In the previous chapter (p. 93) above we quoted from *Master Sūn's Mathematical Manual*: 'There are an unknown number of things. Three by three, two remain; five by five, three remain; seven by seven, two remain. How many things?' We are now concerned with the generalization of this problem (the so-called 'problem of the Master Sūn').

Putting this into modern mathematical notation, if we are counting by multiples of m_i with remainder r_i ($i = 1, 2, 3, \ldots$), then the problem is equivalent to solving the following linear congruences:

$$N \equiv r_1 \quad (\mathrm{mod}\ m_1)$$
$$N \equiv r_2 \quad (\mathrm{mod}\ m_2)$$
$$N \equiv r_3 \quad (\mathrm{mod}\ m_3)$$
$$\cdots \qquad \cdots$$
$$\cdots \qquad \cdots$$
$$\cdots \qquad \cdots$$

That is, to find the smallest positive integral value of N. If all of these m_i are co-prime then finding a series of values of $a_1, a_2, a_3, \ldots$ such that a_i satisfies

$$a_i \frac{M}{m_i} \equiv 1 \pmod{m_i},$$

where $M = m_1.m_2.m_3. \ldots$ (that is, the product of the m_i), obviously shows that

$$N = \left[r_1 a_1 \frac{M}{m_1} + r_2 a_2 \frac{M}{m_2} + r_3 a_3 \frac{M}{m_3} + \ldots \right] - \theta M$$

is a solution where θ is an integer. By choosing θ suitably we can make N the smallest positive integer to satisfy the given conditions.

In an earlier section (see p. 94) it was mentioned that this sort of problem has very close connexions with the calculation of the 'number of years from the initial point (上元積年 , shàng yuán jī nián)' in the computation of the calendar in ancient China. Unfortunately the different schools calculating the calendar for the years from the end of the Hàn Dynasty to the Southern Sòng (220–1127 AD) all just gave the solutions to the 'number of years from the initial position' without describing the method of calculation. According to the extant material the first to give a systematic description of this sort of calculation technique was Qín Jiǔsháo. In Chapters 1 and 2 of Qín Jiǔsháo's *Mathematical Treatise in Nine Sections* (1247 AD) the method of calculation was presented in a systematic way. He also extended this type of calculation of the 'number of years from the initial position' to various kinds of mathematical problem.

From the description above we can see that the key to the whole problem is finding a_i that satisfy

$$a_i \frac{M}{m_i} \equiv 1 \pmod{m_i}.$$

When the given numbers are as simple as the numbers given in 'Master Sūn's problem' then one can find such a_i by trial and error. But when the given numbers and conditions are relatively complicated trial and error is not sufficient. Qín Jiǔsháo introduced a type of method called literally 'the method of finding one by the great extension' (大衍求一術 , dà yǎn qiú yī shù). This method is similar to the present day method of finding the largest common divisor, the so-called Euclidean algorithm. The fact that for any n given numbers m_i, which are co-prime, and any n given numbers r_i, then there does exist a number N as above is still called the 'Chinese remainder theorem' in the West.

The correct explanation of the Chinese name was given by Needham (1959, p. 119 and footnote j). He explains that when one consults the oracle given by the *Yì Jīng* (*I Ching, Book of Changes*) 50 yarrow stalks are used and one of them is put aside before the remainder are divided into two groups signifying the Yīn (陰) and Yáng (陽) (the basic complementary opposites in the cosmos). Needham continues 'it was very natural, therefore, that mathematicians, seeking for remainders of one by continued divisions, should have remembered this'.

Borrowing modern mathematical notation Qín Jiǔsháo's method can be described as follows.

First subtract m_i repeatedly from $\dfrac{M}{m_i}$ so as to make the final result, G, satisfy: $G < m_i$. then at that time G certainly also satisfies

$$G \equiv \frac{M}{m_i} \pmod{m_i}.$$

Qín Jiǔsháo laid out the configuration as in the diagram below, with G in the top right, m_i at the bottom right, 1 at the top left and the bottom left remaining empty. Calculations then proceed in the registers that contain four counting rod numbers.

$$\begin{array}{|cc|} \hline 1 & G \\ 0 & m_i \\ \hline \end{array}$$

The method of calculation in Qín Jiǔsháo's original text for the 'method of finding one by the great extension' says:

First divide right bottom by right top [divide m_i by G], multiply the quotient obtained [Q_1] by the top left [1] and enter in the bottom left [add it into the bottom left, at the same time replacing m_i by the remainder R_i of dividing G by m_i]. And then use the right column top and bottom; using the smaller to divide the greater, dividing alternately [dividing alternately and at the same time continually replacing the old remainder by the new remainder], immediately multiply by the quotient obtained successively entering into the left column top or bottom until finally the top right is just 1, then stop [continue calculating until the top right is 1, then stop]. Then take the top left result and use it as the multiplication ratio [that is a_i].

Let successive divisions have produced quotients $Q_1, Q_2, Q_3, \ldots, Q_n$ and remainders $R_1, R_2, R_3, \ldots, R_n$ and let the top and bottom numbers calculated for the left column be $k_1, k_2, k_3, \ldots, k_n$. Then the figures below show the four registers in the counting rod configurations left and right, top and bottom, and how they change.

$$\begin{array}{|cc|}\hline 1 & G \\ 0 & m_i \\ \hline\end{array} \quad \begin{array}{|cc|}\hline 1 & G \\ k_1 & R_1 \\ \hline\end{array} \quad \begin{array}{|cc|}\hline k_2 & R_2 \\ k_1 & R_1 \\ \hline\end{array} \quad \begin{array}{|cc|}\hline k_2 & R_2 \\ k_3 & R_3 \\ \hline\end{array} \quad \cdots \quad \begin{array}{|cc|}\hline k_n & R_n \\ k_{n-1} & R_{n-1} \\ \hline\end{array}$$

The changes in the two columns can be recorded using modern mathematical formulae by the following two sequences:

$$m_i = GQ_1 + R_1, \qquad k_1 = Q_1;$$
$$G = R_1Q_2 + R_2, \qquad k_2 = Q_2k_1 + 1;$$
$$R_1 = R_2Q_3 + R_3, \qquad k_3 = Q_3k_2 + k_1;$$
$$R_2 = R_3Q_4 + R_4, \qquad k_4 = Q_4k_3 + k_2;$$
$$R_3 = R_4Q_5 + R_5, \qquad k_5 = Q_5j_4 + k_3;$$

$R_{n-2} = R_{n-1}Q_n + R_n (R_n = 1)$, $k_n = Q_nk_{n-1} + k_{n-2}$ is always even. If $R_{n-1} = 1$ and $n - 1$ is odd, then Qín Jiǔsháo uses 1 to divide R_{n-1} so that $R_{n-1} = 1.0 + R_n$, the quotient is zero and the nth remainder, $R_n = 1$. Solving the congruence

$$a_1.2970 \equiv 1 \pmod{83}$$

(which is Problem 1 of Chapter 2 of the *Mathematical Treatise in Nine Sections*) as a specific example will give an overall understanding of the above mentioned two sequences of formulae. This time $G = 2970 - 35.83 = 65$, $m_i = 83$ and the two sequences of formulae are respectively,

$$83 = 1.65 + 18, \quad k_1 = 1;$$
$$65 = 3.18 + 11, \quad k_2 = 3.1 + 1 = 4;$$
$$18 = 1.11 + 7, \quad k_3 = 1.4 + 1 = 5;$$
$$11 = 1.7 + 4, \quad k_4 = 1.5 + 4 = 9;$$
$$7 = 1.4 + 3, \quad k_5 = 1.9 + 5 = 14;$$
$$4 = 1.3 + 1, \quad k_6 = 1.14 + 9 = 23.$$

So finally, since $R_6 = 1$ and $k_6 = 23$, 23 is the required a_1. We can readily verify that $23.2970 \equiv 1 \pmod{83}$.

When we need to find $k_n = a_i$ satisfying the congruence

$$k_n.\frac{M}{m_i} \equiv 1 \pmod{m_i}$$

the general proof is also not difficult. Let $\ell_2 = Q_2$, $\ell_3 = Q_3\ell_2 + 1$, $\ell_4 = Q_4\ell_3 + \ell_2, \ldots, \ell_n = Q_n\ell_{n-1} + \ell_{n-2}$, then from the two sequences of formulae above one can calculate that:

$$R_1 = m_i - k_1G,$$
$$R_2 = G - Q_2R_1 = G - Q_2(m_i - k_1G)$$
$$= k_2G - \ell_2m_i,$$
$$R_3 = R_1 - Q_3R_2 = (m_i - k_1G) - Q_3(k_2G - \ell_2m_i)$$
$$= \ell_3m_i - k_3G,$$

. . .

$$R_{n-1} = \ell_{n-1}m_i - k_{n-1}G,$$
$$R_n = k_nG - \ell_nm_i.$$

When $R_n = 1$ then the last equation is $k_n G - \ell_n m_i = 1$, which allows us to deduce that

$$k_n . G \equiv 1 \pmod{m_i}$$

and since we know G also satisfies $G \equiv \dfrac{M}{m_i}$ we therefore know that

$$k_n . \frac{M}{m_i} \equiv 1 \pmod{m_i}$$

Because in finding k_n we go on until the top right becomes 'just 1, then stop,' Qin Jiǔsháo called this the 'technique of finding one'. He further combined this calculation procedure with the 'great extension number' (大衍之數, Dà yǎn zhī shù) of the *Great Appendix to the Book of Changes* (易經 · 繫辭傳 Yì jīng Xì cí zhuàn) and called it the 'method of finding one by the great extension'.

In the computation of the calendar, as mentioned previously, the m_i are the periods of the motions of the various heavenly bodies (such as the period of the orbit, the phases of the moon . . .) so the m_i cannot be integral. In his book Qin Jiǔsháo researched into four possibilities for the m_i classified as 'natural numbers' (元數, yuán shù) 'converging numbers' (收數, shōu shù) 'common numbers' (通數, tōng shù) and 'repeating numbers' (復數, fù shù). 'Natural numbers' are ordinary positive integers; 'converging numbers' are decimal fractions; 'common numbers' are rational numbers and 'repeating numbers' are numbers ending in zeroes, that is, multiples of powers of ten. When one of these situations is encountered Qin Jiǔsháo always converted it into a situation of the first type before proceeding to calculate.

When the m_i are not co-prime Qin Jiǔsháo employs procedures such as 'cancel not one place but in many places' and 'cancel the even but not the odd'. Although the concepts of prime number and of prime factor of a positive integer had not been introduced at that time, nevertheless Qin Jiǔsháo had specified methods of calculation to make up for deficiencies in these respects.

Only the nearest distances and not the 'accumulated years from the initial point' (上元積年, Shàng yuán jī nián) were calculated in the *Works and Days Calendar* of the Yuán Dynasty. The Dàtǒng Calendar (大統曆, Dà tǒng lì) decreed by the Míng Dynasty (1368–1644 AD) is basically drawn from the *Works and Days Calendar* but also disregarded the method of calculating the 'accumulated years from the initial point'. This can be said to be some sort of improvement in the computation of the calendar but from then on the 'method of finding one by the great extension', which owes its origins and development to the requirements of calculating the calendar was gradually also lost.

In the middle of the Qīng Dynasty a lot of mathematicians researched into ancient mathematics and rediscovered this method of calculation.

In Europe it was only in the eighteenth and nineteenth centuries that Euler

(1707–1789 AD), Gauss (1777–1855 AD) and others researched into linear congruences. In this regard Qín Jiǔsháo's research was more than 500 years earlier than the work in Europe.

2 Magic squares

This is a relatively exotic component of the mathematical texts of the Sòng and Yuán period.

The 'row and column diagram' (縱橫圖, zòng héng tú) (that is, the modern day magic square) has remote origins in ancient China. Earlier when we introduced the *Memoir on Some Traditions of Mathematical Art* (數術記遺, *Shùshù jiyí*) we mentioned a 3 × 3 square: the 'nine houses diagram'. Yáng Huī in his book *Continuation of Ancient Mathematical Methods for Elucidating the Strange [Properties of Numbers]* (續古摘奇算法, *Xùgǔ zhāiqi suànfǎ* — 1275 AD) went further and listed magic squares up to 10 × 10. In Fig. 5.3 we present for examination a 7 × 7 square, [Yáng Huī called this the 'extension number diagram' (衍數圖, yàn shù tú) for the 'extension number' (衍數, yàn shù), as mentioned in the Yì Jīng and above in the preceding section, is 49], a 9 × 9 square and a 10 × 10 square. (Notice that the sums of the numbers in any row, column or main diagonal are all the same, apart from the 10 × 10 square, where the diagonal is excepted.)

It is worth noting that like the examples in Fig. 5.3, the 7 × 7 magic square, that is the so-called 'central magic square' (中心幻方, Zhōng xīn huàn fāng), was passed in through the western boundary (Turkestan) during the Yuán period and in 1956 a 6 × 6 magic square inscribed on an iron tablet was discovered in Xī'ān (西安). This is also commonly known as a 'central magic square' (details in the next section).

A magic square uses n^2 successive natural numbers placed in n^2 squares so that the sums of the various numbers along any column, row or main diagonal are the same, which of course, requires a great deal of skill. However at the beginning of the Northern Sòng Dynasty people took the 3 × 3 magic square, which forms the so-called 洛書 (luò-shū literally 'Luò river writing') diagram from the *Yì Jīng* and conjured up a lot of mysticism from it. They thought that the origin of mathematics was connected with 'luò-shū', but this is completely erroneous. Some of the mathematical works of the Míng and Qīng periods (1368–1912 AD), such as the *Systematic Treatise on Arithmetic* (算法統宗, *Suàn fǎ tǒngzóng*) of 1593 and the *Collected Basic Principles of Mathematics* (數理精蘊, *Shù lǐ jīng yùn*) of 1723, have erroneously taken the 'luò-shū' as the source of mathematics. (For a historical account of the luò-shū and hétú 河圖, literally 'river diagram', see Needham 1959, pp. 56–9.)

(a)

46	8	16	20	29	7	49
3	40	35	36	18	41	2
44	12	33	23	19	38	6
28	26	11	25	39	24	22
5	37	31	27	17	13	45
48	9	15	14	32	10	47
1	43	34	30	21	42	4

(b)

31	76	13	36	81	18	29	74	11
22	40	58	27	45	63	20	38	56
67	4	49	72	9	54	65	2	47
30	75	12	32	77	14	34	79	16
21	39	57	23	41	59	25	43	61
66	3	48	68	5	50	70	7	52
35	80	17	28	73	10	33	78	15
26	44	62	19	37	55	24	42	60
71	8	53	64	1	46	69	6	51

(c)

1	20	21	40	41	60	61	80	81	100
99	82	79	62	59	42	39	22	19	2
3	18	23	38	43	58	63	78	83	98
97	84	77	64	57	44	37	24	17	4
5	16	25	36	45	56	65	76	85	96
95	86	75	66	55	46	35	26	15	6
14	7	34	27	54	47	74	67	94	87
88	93	68	73	48	53	28	33	8	13
12	9	32	29	52	49	72	69	92	89
91	90	71	70	51	50	31	30	11	10

Fig. 5.3 (a) 'Extension number diagram' (7 × 7 magic square). Sum of any row, column or diagonal is 75. (b) A 9 × 9 magic square. Sum of any row, column or diagonal is 369. (c) (百子圖, Bǎizǐ tú — 10 × 10 magic square). Sum of any row or column is 505.

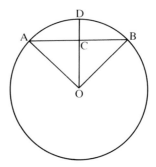

Fig. 5.4 Cutting the circle diagram.

3 The ideas on spherical trigonometry in the *Works and Days Calendar*

Astronomers in ancient China, India, and the Arab world used spherical trigonometry for performing calculations in astronomy from an early stage. During the Suí and Táng periods (589–907 AD) Indian astronomy passed into China, but its sorts of method did not arouse the interest of Chinese astronomers and mathematicians.

There is a brief mention of the relationship between the chord, sagitta, and arc in the ancient Chinese text *Nine Chapters on the Mathematical Art*. For example in the *Gōugǔ* chapter there is a passage: 'A circular wooden pillar is embedded in a wall. Its size is unknown. Using a saw to saw it, at a depth of one inch the saw length is one foot. What is the diameter?' and similar problems. The 11th century scientist Shĕn Kuò (沈括) of the Sòng Dynasty introduced a formula relating the chord, sagitta, and arc. As in Fig. 5.4 let the diameter of the circle be d, the radius r, the chord AB $= c$, the sagitta CD $= v$, and the arc $ADB = s$. Then in modern mathematical notation Shĕn Kuò's formulae are equivalent to

$$\begin{cases} \dfrac{c}{2} = \sqrt{(r^2 - (r-v)^2} = \sqrt{(dv - v^2)} \\ s = c + \dfrac{2v^2}{d}. \end{cases}$$

Shĕn Kuò called this method of calculation the 'technique of intersecting circles' (會圓術, Huì yuán shù). We know that the second formula is only approximate.

Shĕn Kuò's 'technique of intersecting circles' was used by Wáng Xún (王恂), Guō Shŏujìng (郭守敬) *et al.* in the calculations of 'accumulated degrees along the equator' (latitude) and 'inclination to the equator' (longtitude). In the process of calculation they also derived some relationships. Discovering these relationships is tantamount to opening the door into spherical trigonometry.

Finding the degrees of latitude and longitude knowing the position of the sun in terms of its angular position in the ecliptic is called calculating the 'accumulated degrees along the equator' and the 'inclination to the equator' [literally, the in and out degrees]. In Fig. 5.5 arc AD is a quarter of the ecliptic circle, arc AE is its projection on the equator, and arc DE is called the 'great distance between the ecliptic and the equator'. Suppose the sun moves along the arc AD to a point B, then arc BD is the 'accumulated degrees along the ecliptic', arc CE is the 'accumulated degrees along the equator', and arc CB is the 'inclination to the equator'. (F is the south pole.)

We first take the relations in the calculation of the 'great distance between the ecliptic and the equator' as an example, i.e. the relations in the calculation of the sagitta KE with respect to the arc ED, to illustrate the use of the technique of intersecting circles. Let arc DE = s, the diameter of the sphere (2OE) = d, the radius (OE) = r, the sagitta KE = v (this is equal to the central sagitta), DK = p (this is equivalent to the gōu), and OK = q (this is equivalent to the gǔ). Then from the formulae of the 'technique of intersecting circles' we have:

$$\begin{cases} p = \sqrt{[r^2 - (r - v)^2]} = dv - v^2 \\ s = p + \dfrac{v^2}{d}. \end{cases}$$

Eliminating p we get $v^4 + (d^2 - 2ds)v^2 - d^3v + d^2s^2 = 0$. Solving we obtain v, and substituting in $q = r - v$ and $p = \sqrt{(dv - v^2)}$ we obtain p and q.

Finding the 'accumulated degrees along the equator' and the 'inclination to the equator' in the *Works and Days Calendar* is equivalent to finding the number of degrees in the arcs CE and CB, given the arc BD in degrees.

Referring to Fig. 5.5, from B construct BL $\perp$ OD, then using the 'tech-

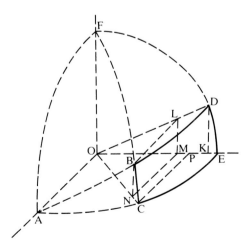

Fig. 5.5 The 'accumulated degrees along the equator' and the 'inclination to the equator'.

nique of intersecting circles' we similarly obtain: sagitta LD $= v_1$ in terms of the half-chord LB $= p_1$, and gǔ OL $= q_1$.

Again from L construct LM $\perp$ OE, from B construct BN $\perp$ OC, join MN, then we know MN $=$ LB $= p_1$. Let the half cord BN of the arc BC be p_2, the gǔ ON $= q_2$ and the sagitta NC $= v_2$. Then because $\triangle$OML is similar to $\triangle$OKD, we obtain BN $=$ LM $= \dfrac{\text{OL}}{\text{OD}}$. DK, that is

$$p_2 = \frac{q_1 p}{r} \tag{5.9}$$

Again, knowing OM $= \dfrac{\text{OL}}{\text{OD}}$. OK and ON $= \sqrt{(\text{OM}^2 + \text{MN}^2)}$ (since $\angle$ OMN is a right angle) we arrive at

$$q^2 = \sqrt{\left[\left(\frac{q_1 q}{r}\right)^2 + p_1^2\right]}.$$

NC $=$ OC $-$ ON, that is

$$v_2 = r - q_2.$$

From the sagitta v_2 and the half chord p_2 of the arc BC the 'technique of intersecting circles' allows us to derive:

arc BC $=$ 'inclination to the equator' when the sun

is over point B $= p_2 + \dfrac{v_2^2}{d}$.

The method for deriving the 'accumulated degrees along the equator' is quite similar. Let CP $>$ OE, then $\triangle$OPC is similar to $\triangle$OMN.

Let the half chord of the arc CE, CP $= p_3$, the gǔ OP $= q_3$, and the sagitta PE $= v_3$. Then since CP $= \dfrac{\text{OC}}{\text{ON}}$.MN we can obtain

$$p_3 = \frac{r p_1}{\sqrt{\left[\left(\frac{q q_1}{r}\right)^2 + p_1^2\right]}}. \tag{5.10}$$

Again since OP $= \dfrac{\text{OC}}{\text{ON}}$.OM, we can obtain

$$q_3 = \frac{q q_1}{\sqrt{\left[\left(\frac{q q_1}{r}\right)^2 + p_1^2\right]}}. \tag{5.11}$$

Again PE $=$ OE $-$ OP, so we get

$$v_3 = r - q_3.$$

From p_3 and v_3 the 'technique of intersecting circles' then gives:

arc CE = 'accumulated degrees along the equator' when the sun is
　　　　over point B

$$= p_3 + \frac{v_3^2}{d}.$$

In fact the method for obtaining the 'inclination to the equator' and the 'accumulated degrees along the equator' is similar to the method of solving right-angled triangles in spherical trigonometry. If c represents the angle measured in radians of the arc AB of the angular distance along the ecliptic, b represents the angular distance measured in radians of the arc AC, a the angular distance in radians of the arc CB, and α the angle EOD at the point of intersection of the planes of the ecliptic and the equator, then dividing both sides of eqns (5.9), (5.10), and (5.11) by radius r we get the following formulae of spherical trigonometry:

$$\sin a = \sin c \sin \alpha$$

$$\cos b = \frac{\cos c}{\sqrt{(\sin^2 c \cos^2\alpha + \cos^2 c)}}$$

$$\sin b = \frac{\sin c \cos \alpha}{\sqrt{(\sin^2 c \cos^2\alpha + \cos^2 c)}}.$$

Although Wáng Xún, Guō Shǒujìng and others introduced this new method of calculation, the 'technique of intersecting circles' used $\pi = 3$, which involves a large error, and so the results calculated are not very accurate. Although they introduced the new sort of spherical trigonometry this new method did not continue to develop. The universal adoption of the methods of spherical trigonometry in China had to wait until as late as the 17th century when mathematics came in from the West (details below).

5.6 The interchange of mathematical knowledge between China and the outside world in the Sòng and Yuán period

In the Sòng and Yuán periods (960–1280 AD), the interchange of mathematical knowledge between China and the outside world greatly improved over the situation in the Hàn to the Táng (206–907 AD). In particular, in the 13th century, following the western advances of the Mongolians, there were remarkable developments in cultural exchange between China and the Islamic countries. In the year 1271 AD Khublai Khan, the founder of the Yuàn dynasty, decreed the establishment of an Islamic observatory in the capital. The astronomer Jamāl al-Dīn from Persia worked in this institute. He once made some astronomical instruments and wrote a book on the Islamic calendar — *The Perpetual Calendar* (萬年曆, *Wàn nián lì*). It is recorded in Chapter 7 of the *Collection of Official Records of the Yuán dynasty* (元秘書監志 , *Yuán bì shū jiān zhì*), in the section on Islamic books

(回回書籍, Huí huí shū jí), that at that time some astronomical texts passed from the Islamic countries into China. These works include the *Four Cuts Methods for Periodic Numbers of Wùhūlǐ* (兀忽列的四擘算法段數 , *Wùhūlǐde sìbì suànfǎ duàn shù*) in 15 chapters; the *Method of Appropriate Solution of Hǎnlǐliánkū* (罕里連窟允解算法段目 , Hǎnlǐliánkū yǔn jiě suànfǎ duàn mù), three chapters; *Various Methods and Forms of Sāwéinàhǎndāxiyá* (撒唯那罕答昔牙諸般算法段目並儀式 *Sāwéinàhǎndāxiyá zhū bān suànfǎ duàn mù bìng yí shì*), 17 chapters; and *Various Methods of Calculation of Axiēbiyá* (呵些必牙諸般算法 , *Axiēbiyá zhū bān suànfǎ*), eight chapters. These books were not translated into Chinese and the original texts have long been lost. However it is possible that 'Wùhūlǐ' is a transcription of 'Euclid' and that the book was transcribed into Chinese from an Arabic version of Euclid. The names in the other titles may also be transcriptions from Arabic into Chinese (cf. Needham 1959, p. 105).

At that time Arabic numerals also passed into China. In excavations in the winter of 1956 on the outskirts of Xī'ān, which is a former site of the residence of the Ān Xī Prince (安西王府, Ān Xī wáng fǔ), five iron tablets were discovered and on them were inscribed 6 × 6 magic squares as shown in Fig. 5.6. According to the superstitious practice of those days the iron tablets were used for deterring evil spirits and for exorcism. The figures inscribed are very similar to the so-called East Arabic numerals.

In those days the Arab world used the 'method of calculation on sand', and this too passed into China. This mode of calculation uses bamboo or iron rods on the 'sand board or earth board (筆算, bǐ suàn)'. A poem by the Míng Scholar Táng Shùnzǐ (唐順之) says: 'Translating the western barbarian calendar by sand script'; here 'sand script' indicates this sort of calculation on a sand board.

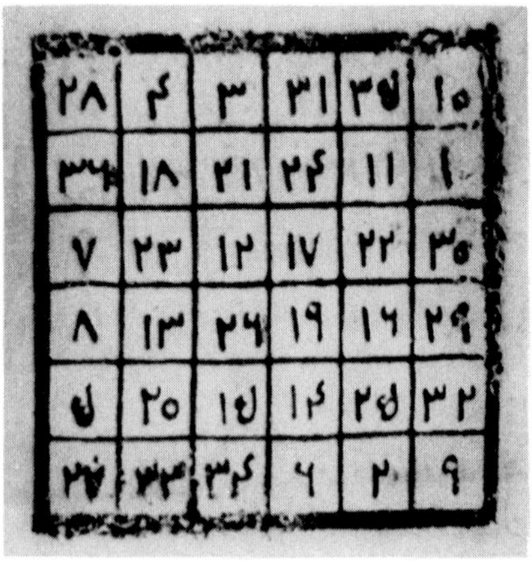

28	4	3	31	35	10
36	18	21	24	11	1
7	23	12	17	22	30
8	13	26	19	16	29
5	20	15	14	25	32
27	33	34	6	2	9

Fig. 5.6 The iron plate unearthed at Xī'ān. Left: print of the East Arabic numeral plate (one quarter of the size of the original). Right: in modern Hindu-Arabic numerals.

Fig. 5.7 Wǔ Jìng's 'written calculation'.

The Míng mathematician Wǔ Jìng (吳敬) introduced a method of 'written calculation' (寫算, xiě suàn) in his work. Wǔ Jìng said: 'In written calculation, first display a grid, place the multiplier in the top row and the multiplicand in the right column, bring the multiplicand and the multiplier together and fill in the squares . . .' Figure 5.7 shows the calculation of 'knowing the price of one shí (石) [weight] of cereal is 45 guàn (貫) and 678.9 copper cash [1 guàn = 1000 copper cash]. Now [we] have 425 shí of cereal. What is the total price?' The written calculation is shown. Chéng Dàwèi (程大位) in his *Systematic Treatise on Arithmetic* (1592 AD) towards the end of the Míng Dynasty, called this method of calculation 'scattering on the ground' (鋪地錦, pū dì jǐn). This mode of calculation also came from the Arab world.

During the time when mathematical knowledge passed from the Arab world of the West into China, Chinese mathematical knowledge was also exchanged. After the Mongolians occupied Baghdad the observatory Maraghah was built there on the suggestion of the astronomer and mathematician Nasir al-Dīn al-Tusī. This observatory acquired famous scholars from all over. Amongst them were four astronomers that the Mongolian commander Hulagu Khan brought with him at that time from China, and of course these astronomers were well versed in Chinese mathematics.

Some problems used in some of the mathematical writings of the Islamic countries to the west of China were very similar to problems in the ancient mathematical books of China. For example the problems of 'knowing the number of objects', solving congruences in the *Mathematical Manual of Master Sūn*, and the 'problem of a hundred fowls' in *Zhāng Qiūjiàn's Mathematical Manual*. The description of the methods of division, extracting square roots and cube roots, and other techniques in the *Key to Arithmetic* by the mathematician al-Kāshī (Miftāḥ al-Ḥisāb) (1427 AD) are also very similar to the methods used in ancient China (for example the method of 'moving position' in extracting square roots). In particular the method of extraction of higher roots and the table of binomial coefficients i.e. Jiǎ Xiàn's diagram on the source of the 'method of extracting roots' (Pascal's triangle), used when calculating approximate roots in the procedure for extracting higher degree roots is exactly the same as in the mathematical texts of the Sòng and Yuán period in China. Apart from the records in the Chinese texts of the Sòng and Yuán period this sort of mathematical knowledge had never been mentioned previously in any form in any country.

Besides being propagated to the West, Chinese mathematical texts had tremendous influence in Korea and Japan to the East. The Koreans reprinted Yáng Huī's *Mathematical Manual*, Zhū Shìjié's *Introduction to Mathematical Studies* and other works. The *Introduction to Arithmetic* was lost in China and the copies that survive to the present depend for their reappearance on the reprinting of the Korean edition. In Japan besides reprintings, various commentaries were written.

6

The evolution from calculating with counting rods to calculating with the abacus

6.1 The background situation during the beginnings of abacus calculation

In the previous chapter we reported the tremendous developments in mathematics in ancient China during the Sòng and Yuán period. These achievements were extensions based on counting rod calculations. However it must be noted that this tremendous development did not last, it did not progress to a higher level; just the opposite, many important achievements such as the 'method of the celestial element', the 'method of four unknowns', the summation of higher order differences series, rapidly declined and in some cases were in danger of being lost. By the 15th century some Míng Dynasty mathematicians had virtually no understanding of the 'method of the celestial element' or the 'method of four unknowns'.

How did this state of affairs arise? There are many reasons for it. Most importantly, the sources of development were detached from the actual requirements of Chinese society at that time. It was indeed very rare, for example, for the 'method of the celestial element', to find a practical application. There is some use of the 'method of the celestial element' in the book *On the Prevention of River Flooding* (河防通議 , *Hé fáng tōng yì*, 1321 AD) by the minority race scientist Shā kèshí (沙克什), and in the *Works and Days Calendar* (授時曆 , Shòu Shí Lì, 1280 AD) by Wáng Xún (王恂), Guō Shǒujing, and others the 'method of the celestial element' was also used. [N.B. Needham (1959) uses the title *Shou Shih Calendar* for this calendar.] However, only quadratic equations were used in the *On the Prevention of River Flooding* and after the completion of the *Works and Days Calendar* the officials in the State Observatory in the Míng period only knew how to use the tables in the *Works and Days Calendar* for their calculations and did not know how they were computed. Most of the problems in Lǐ Yě's texts *Sea Mirror of Circle Measurements* and *New Steps in Computation* are problems made up for answers already known. We cannot find any factual, relevant problems when we look for them in the light of the social and economic requirements of society at that time, even including the requirements for the development of science. Problems in those days which required the solving of quadratic or higher degree equations did not come from the actual requirements of production and livelihood. Further, no kind of factual, relevant examples involving systems of higher degree equations can be found.

As far as problems on finding the sums of higher order equal difference series and interpolations are concerned the situation is quite similar.

In addition to the detachment of mathematical knowledge from the practical requirements of the society of the time, the content of much of this mathematics was difficult and hard to understand so much mathematical knowledge was lost. For example most of the problems involving the method of the celestial element and the method of four unknowns were formulated in terms of conditions on relations between line segments in right-angled triangles. This deficiency decided the fate of the method of four unknowns and prevented it from developing readily and continuously.

What were the actual needs of society at that time? What questions arose for mathematics in those days produced by the society?

During the 250 or more years of the Ming Dynasty, from its establishment to its demise in the 16th century, the overall trend in the development of the economy of the society was upward, although there were fluctuations. There were great developments in various types of handicraft, commerce, and overseas trade. These sorts of development, in particular the developments in commerce and trade, raised problems for mathematicians in carrying out the computations for dealing with the ever-increasing size and complexity of data. Most computational problems (mainly addition, subtraction, multiplication, and division — the four arithmetic calculations) demanded faster calculation and greater convenience. Under the circumstances the techniques of counting rods passed down from the ancients became inadequate in the following respects. First, the operations of multiplication and division require a display of three rows from top to bottom in counting rod calculations and this is not very convenient. Second, in counting rod calculations the techniques for addition and subtraction demand continual changes in the counting rod configurations. This leads to difficulty when doing calculations quickly. Although a start was made on using written computations to replace counting rod displays in the Sòng period, the basic features of counting rod calculation techniques were unchanged.

As a consequence, a new form of calculating with the abacus was born to adapt to the new situation.

The appearance of the abacus was a major event in the history of Chinese mathematics. The calculation device was changed completely. This easy-to-carry and simple calculating device is still widely used throughout China. At the time of its invention, the abacus passed over to Korea, Japan, and other Asian countries, where it is still used to the present day. Abacus calculation has had a big influence both within and outside China.

Abacus calculation developed from the basic calculations with counting rods. So what were the conditions required for counting rod calculations to evolve into abacus calculations? And what is the general picture of this evolution? We shall briefly describe it below in three stages.

6.2 The simple and quick techniques for multiplication and division from the end of the Táng Dynasty

The revolution in counting rod calculations had already started as early as about the eighth century AD, towards the end of the Táng period. These changes cannot be divorced from the economic developments in the middle Táng society. They principally involved the improvement of techniques for multiplication and division.

As mentioned in earlier sections, the methods of multiplication and division in the counting rod calculations of ancient China required displaying three rows: top, middle, and bottom. Mathematicians at the end of the Táng worked hard to find a 'faster method for multiplication and division' so that the three row configuration for multiplication and division could be effected using just one row of counting rods.

There are some examples preserved in *Xiàhóu Yáng's Mathematical Manual*, which is still extant. (As mentioned previously, the surviving edition of *Xiàhóu Yáng's Mathematical Manual* dates from the late Táng Dynasty — the middle of the eighth century AD.) These examples all make use of the change from many row multiplication and division to single row multiplication and division. They are very good examples for demonstrating the change from three row calculation to single row calculation. For example when the multiplier is 35, 5 is used to multiply first followed by 7; when the divisor is 12, first halve and then divide by 6; when converting weights in jīn (斤) into liǎng (兩) (1 jīn = 16 liǎng), multiply first by 2 and then multiply by 8; when converting from liǎng to zhū (銖) (1 liǎng = 24 zhū), multiply first by 3 and then by 8.

When the leading digit of the multiplier is 1 it is possible to 'use addition instead of multiplication'. For instance to calculate when the multiplier is 14, one uses the method of 'adding to the given fourfold'; when the multiplier is 144, one uses the method of 'adding to the given forty fourfold'; when the multiplier is 102, one uses 'skip a place and add on double'. Similarly, in division, when the leading digit of the divisor is 1 it is possible to use the method of subtraction instead of division. For example, when the divisor is 12 one can proceed to calculate by using the method of 'subtracting from the given twofold'.

When the leading digit of the multiplier or divisor is not 1 in a multiplication or division, various methods can be used to transform the leading digit so as to convert it into '1' for calculating. This sort of 'transforming into 1' technique is called 'finding 1' (得一 , Dé yī) or 'deriving 1' (求一 , Qiú yī). In the section on 'Art and craft' in the *New History of the Táng Dynasty* (新唐書 · 藝文志 , *Xīntáng shū Yì wén zhi*) there is reference to 'Jiāng Běn's (江本) *Techniques of Calculating a Leading 1* (一位算法 , Yī wèi suàn fǎ) in two chapters and Chén Cóngyùn's (陳從運) *Mathematical Manual on Deriving 1*, 得一算經 , Déyī suànjīng) in seven chapters. These books are long lost. In the volume 'Memoir on the Calendar' in the *Sòng History* (宋史 · 律曆志 , *Sòng shū lù lì zhì*) it reports: 'Chén Cóngyùn wrote the *Mathematical Manual on Deriving 1*. Its techniques are based on doubling and halving, and adopting the principles of subtracting and adding so as to transform into forms suitable for calculation.' These are exactly the central ideas of the calculation methods of 'finding 1' or 'deriving 1'.

Shěn Kuò, the Northern Sòng period scientist, says in his *Dream Pool Essays*: 'in arithmetic seeing simplicity use it, seeing complexity change it, don't stick to one method' and this already clearly points the way to revolutionizing counting rod calculations; the trend should be to convert the complex into the simple.

These types of change in the procedures for multiplying and dividing in counting rod calculations continued increasingly throughout the Sòng and Yuán period. There are some items in the extant works of Yáng Huī of the Southern Sòng (see Lam Lay-Yong 1977, Chapter 2). For example, in his work *Precious Reckoner for Variations of Multiplication and Division* (乘除通變算寶 , *Chéng chú tōng biàn suàn bǎo*) there is a systematic description of the method of 'using addition instead of multiplication and using subtraction instead of division'. Besides the description in the book of 'adding one more position' (the multiplier being 11, 12, . . ., 19), 'adding two more positions' (the multiplier being 111, 112, . . ., 199), 'repeated addition' (the multiplier being factored, e.g. 247 = 13 × 19), 'addition skipping a place' (101, 102, . . ., 109), etc., he uses the methods of 'subtracting one more place', 'subtracting two more places', 'repeated subtraction', 'subtraction skipping a place' and besides all these he also composed eight lines on the method of 'replacing multiplication by converting to 1'. One can rapidly grasp the method of calculation of 'replacing multiplication by converting to 1' through these eight lines. The eight lines run: '5, 6, 7, 8, 9, double the multiplier. 2, 3 need halving, encountering 4 halve it twice. Doubling or halving the multiplier, do the opposite to the multiplicand. Use addition to replace multiplication, then the product can be found.' In the two lines 'Doubling or halving the multiplier, do the opposite to the multiplicand' the fǎ (法) is the multiplier and the shí (實) is the multiplicand. Yáng Huī remarked on these two lines: 'Doubling the multiplier requires halving the multiplicand, halving the multiplier requires doubling the multiplicand.' That is to say if the fǎ is multiplied by 2, the shí requires halving and conversely. The meaning of the complete rhymed formula is: When the leading digit of the multiplier is 5, 6, 7, 8, or 9, it can be doubled so as to make the leading digit 1. When the leading digit of the multiplier is 2 or 3 the multiplier can be halved so as to turn its first digit into 1. When the leading digit of the multiplier is 4 it can be divided by 2 twice, turning its first digit into 1 too. Then use 'add one more position' or 'add two more positions' or etc. so as to 'use addition to replace multiplication' to calculate. At the same time, if the multiplier is doubled the multiplicand must be halved; and if the multiplier is first halved, then the multiplicand must be doubled. (See also Lam Lay-Yong [1977], pp. 36 and 217.)

Besides the lines on 'deriving 1' for the multiplication technique, Yáng Huī's book also contains lines on 'deriving 1' for the method of division. However, after the introduction of 'coming back division' (歸除, guī chú) in the middle of the fourteenth century the 'deriving 1' technique of division became outmoded.

6.3 Folk mathematics and the composition of rhymed formulas for counting rod calculations during the thirteenth and fourteenth centuries

In the previous section we saw that some rhymed formulae had already begun to appear in Yáng Huī's text. In fact the versification of a lot of methods of calculation was the outstanding feature of everyday mathematics in thirteenth and fourteenth century China. These rhymed formulae have a very important place in the course of the evolution from counting rod to abacus calculation.

In the beginning of the Southern Sòng when Róng Qǐ (榮棨) reprinted the *Nine Chapters on the Mathematical Art*, he said in the preface: 'Since the Jīng Kāng reign (靖康) [the reign of the last emperor of the Northern Sòng], . . . concealing the question and answer to cheat the public or versifying to delude oneself goes against the spirit of benefitting the descendants of the myriad generations, it is a means of making a quick profit.' From these words we can see that from the Southern Sòng on this sort of attitude was quite widespread. Besides Yáng Huī's work mentioned above, versifications of various kinds of method of calculation are contained in the works of Zhū Shìjié, Dīng Jù (丁巨), and Jià Héng (賈亨) of the Yuán dynasty and of Liú Shìlóng (劉仕隆), Wú Jìng (吳敬), and Chéng Dàwèi (程大位) of the Míng dynasty; on occasion they also used poetic forms of presenting various types of mathematical problem. In the section *Questions in verse* (或問歌彖 , Huò mén gē tuán) of Zhū Shìjié's *Precious Mirror of the Four Elements* there are altogether 12 problems presented in the form of poems. Among them there are, for example

Problem 1: Now there is a square pond, with the shape of a square platform, each side being 1 zhàng (丈 = 10 feet) 4 [feet]. Reeds grow vertically on the western bank sticking exactly 30 inches out of the water. On the eastern bank another kind of reed grows exactly one foot out of the water. When the two reeds are made to meet they are exactly level with the water surface. Permit me to ask how to determine these three things [the water depth, the lengths of the two reeds]?

Problem 4: I take along a bottle with some wine in it for an excursion in the spring. On reaching a tavern I double its contents and drink one and nine-tenths dǒu (斗) in the tavern. After passing four taverns the bottom is empty. Permit me to ask how much wine was there at the beginning?

As well as problems in the form of poems, some of the calculation techniques were also presented in the form of verses. For example in a mathematical manual of the Yuán Dynasty by the name of *Explanations of Arithmetic* (詳明算法 , *Xiáng míng suàn fǎ*) there is a verse on 'repairing the fort', which says:

In arithmetic there is a rule for building a wall,
take the upper and lower and halve that.
Multiply by the height and then the length.
Clearly this gives the volume of the fort.
Three times the volume and five as one,
this is the method to find the rammed earth from the loose.
Digging out earth apply four times
the amount of loose earth divided by five.

This poem describes the rule given for Problem 1 in Chapter 5 ('Construction consultations') in the *Nine Chapters* (see Vogel 1968, p. 43). The first part just gives the volume as $\frac{1}{2}(t + b)h\ell$, where t is the top length, b the bottom length, h the height, and ℓ the length. The second part says that to get the volume of rammed earth from loose earth multiply by three and divide by five; to get the volume of earth dug out from rammed earth multiply the volume of dug out earth by four and divide by five.

Among the rhymed formulae on 'measuring fields' there is:

The ancient people measured fields by taking length and width,
always relying on string measures to measure.
Each sort of shape has its own method,
but only the method for square fields can be described easily.
Encountering a slanting or concave
one requires cutting and pasting to make a square.
Then multiply the sides to get the area of the field
and divide by 24 to convert it into mǔ (畝).

(A mǔ is 0.067 ha.)

Among the relatively more important verses we must include the 'converting to decimal' rhyme and the 'verses on division'. A long time ago the weight measures in China used 16 liǎng to one jīn. It was required to find the price of one liǎng of goods given the price per jīn: a problem met very often in ordinary daily life. In the Southern Sòng period Yáng Huī wrote the 'converting to decimal' rhyme in his book *Arithmetic Methods for Daily Use* (日用算法 , *Rì yòng suàn fǎ*, 1262 AD) in order to simplify this sort of calculation:

Finding 1, omit a place 625; finding 2 go back a place 125; finding 3 write 1875; finding 4 change it to 25; finding 5 is 3125; finding 6 liǎng price is 375, finding 7 put 4375; finding 8 change it to 5 a total of eight lines. (See also Lam Lay Yong 1972.) In Zhū Shìjié's *Introduction to Mathematical Studies* (1299 AD) there are 15 lines:

1, go back, 625; 2, stay, 125; 3, stay, 1875; 4, stay, 25; 5, stay, 3125; 6, stay, 375; 7, stay, 4375; 8, stay, a single 5; 9, stay, 5625; 10, stay, 625; 11, stay, 6875; 12, stay, 75; 13, stay, 8125; 14, stay, 875; 15, stay, 9375.

Here the aim is to point out that $\dfrac{1}{16} = 0.0625$, $\dfrac{2}{16} = 0.125$, $\dfrac{3}{16} = 0.1875$,

. . . (See Lam Lay Yong 1979, p. 11.)

With this, knowing the price of a jīn and using a line of the 'converting to decimal' rhyme very easily yields the price of a liǎng. For example, knowing the price of a jīn is 56 copper cash, just recite '5, stay, 3125', '6, stay, 375', then the price of a liǎng is 3.125 + 0.375 = 3.5 copper cash. Again, take, for example, conveting 2 jīn 7 liǎng into 2.4375 jīn. In this way complicated workings for multiplication and division can be replaced by additions. These only require a mechanical sort of calculation without having to worry about other conditions. So up to the present day a lot of people use the saying: 'No matter how many it is, one, go back, 625, just do it.' This 'converting to decimal' rhyme gives a transparent technique for finding the liǎng price from the jīn price.

The 'verses on division' have a very important bearing on the change from counting rods to abacus calculations. In calculating with counting rods division was called 'direct division' (商除 , shāng chú). That is to say, during division it is necessary to do testing and checking in order to find the quotient. With the 'verses on division', direct division can be modified so that, as in the 'converting to decimal' rhyme one has '1, go back, 625', so reciting the verse immediately gives the quotient.

Nowadays, almost every Chinese knows 'For 2, 1 becomes a 5; meeting 2, advance 10,' and so on in the 'up to nine verse on division'.

Below we briefly describe the history of the development of the 'up to nine verse on division'.

In the *Dream Pool Essays* by Shěn Kuò of the Northern Sòng a method of continued addition is described, namely: 'Dividing by 9, increase by 1', 'dividing by 8, increase by 2', etc. This is precisely the forerunner of the verse '9, 1, bottom add 1; 9, 2 bottom add 2'. But when the divisor has several digits the method of 'continued addition' is not very convenient.

In the *Precious Reckoner for Variations of Multiplication and Division*, Yáng Huī treats the quick method of 'division in nine parts' where he added 32 new lines, starting from the basis of the four lines of the so-called 'ancient verses' that were current at that time. The 'ancient verses' are:

When the dividend makes the quotient 10, add the remainder to the number above. One half becomes five, do not err in returning to the fixed place. (Cf. Lam Lay Yong 1977 pp. 40 ff.)

Yáng Huī's 32 new lines are:
  'When the dividend makes the quotient 10':

9 as divisor meeting 9 becomes 10 [that is equivalent to, encountering 9 advance to 10, and similarly]; 8 as divisor meeting 8 becomes 10; 7 as divisor meeting 7 becomes 10; 6 as divisor meeting 6 becomes 10; 5 as divisor meeting 5 becomes 10; 4 as divisor meeting 4 becomes 10; 3 as divisor meeting 3 becomes 10; 2 as divisor meeting 2 becomes 10.

  'Add the remainder to the number above':

9 as divisor, seeing 1, put down 1 [that is equivalent to, for 9 dividing 1 add 1 below and similarly] seeing 2 put down 2, seeing 3 put down 3, seeing 4 put down 4; 8 as divisor, seeing 1 put down 2, seeing 2 put down 4, seeing 3 put down 6; 7 as divisor, seeing 1 put down 3, seeing 2 put down 6, seeing 3 put down 12, that is 9 [it should be 'seeing 3 put down 9', but also 'meeting 7 it becomes 10', so 'seeing 3 put down 12' is in fact '7, 3, 42']; 6 as divisor, seeing 1 put down 4, seeing 2 put down 12, that is 8; 5 as divisor, seeing 1 the quotient is 2 [that is equivalent to, 5 dividing 1 doubles it to 2, and similarly] seeing 2, quotient is 4; 4 as divisor, seeing 1 put down 12, that is 6 [that is, '4, 1, 22']; 3 as divisor, seeing 1 put down 21, that is, 7 [that is '3, 1, 31'].

  'One half becomes 5':

9 a divisor, seeing 45 quotient is 5; 8 as divisor, seeing 4 quotient is 5; 7 as divisor, seeing 35 quotient is 5; 6 as divisor, seeing 3 quotient is 5; 5 as divisor, seeing 25 quotient is 5; 4 as divisor, seeing 2 quotient is 5; 3 as divisor, seeing 15 quotient is 5; 2 as divisor, seeing 1 quotient is 5.

  'Do not err in moving back the fixed place-value':

In direct division the dàn (石) place is fixed from the dǒu (斗), now fix the position of the dǒu above the dàn; in direct division getting the quantity in wén (文) from the number of people; now fix the tens position from the number of people [that is: after carrying out the calculation on the dividendum according to the above lines the result moves back one place giving the quotient].

Although Yáng Huī's verses are not exactly the same as the 'division in nine parts' verses of a later date, nevertheless there are a lot of similarities. These verses are still applicable in division when the divisor has only one digit. He invented special sorts of verses for when a divisor has two digits. For example when the divisor is 83 his verses for '83 as divisor' are: 'seeing 1 put down 17, seeing 2 put down 34, seeing 3 put down 51, . . .': Yáng Huī called this sort of division by two-digit divisors 'carrying through division' (穿除 , chūan chú) or 'flying division' (飛歸 , fēi guī).

Zhū Shijié of the Yuán dynasty records 36 lines of 'verses on division' in the *Introduction to Mathematical Studies* (cf. Lam Lay Yong 1979, p. 23).

1 as divisor take as 1, seeing 1 make it 10. 2 dividing 1 increase to 5, seeing 2 make it 10. 3,1,31; 3,2,62; seeing 3 make it 10. 4,1,22;. . . . 9 as divisor, rewrite the

dividendum [that is: 9 dividing 1 put down 1, 9 dividing 2 put down 2, . . . etc.], seeing 9 make it 10.

These lines are exactly the same as the lines for 'division in nine parts' commonly used nowadays in abacus calculations in China.

In those days 'carrying through division' or 'flying division' was not in use in divisions with two-digit divisors; no matter how many digits are in the divisor one can work with the leading digit of the divisor according to the 'division in nine parts' lines in order to find all the digits of the quotient (e.g. the divisor 325 uses '3 as divisor', 4267 uses '4 as divisor') and after obtaining each digit of the quotient use the remaining digits in the divisor to multiply the quotient and then subtract from the dividendum. This kind of division is, in all its essentials, exactly the same as division in present day Chinese abacus calculations except that in those days it was still done using counting rods.

Besides the above mentioned 'division in nine parts' verses there were verses such as 'meeting 1 but does not divide, change to 9 divide 1' in the 'testing the divisor verse' and '1 as divisor, borrow 1 getting 1; 2 as divisor, borrow 2 getting 2' in the 'starting from 1 verse' for the method of verses on division in abacus calculation. These sorts of verse do not appear in Zhū Shìjié's *Introduction to Mathematical Studies*. In the middle of the 14th century lines such as '2 as testing divisor, gives 92, 3 as testing divisor, gives 93' first appeared in Dīng Jù's *Arithmetical Methods* (丁巨算法 , *Dīng Jù suàn fǎ*, 1355 AD). Lines such as 'meeting 2 but does not divide, change to 92, meeting 3 but does not divide, change to 93 . . .' then appeared in the *Explanations of Arithmetic* (詳明算法 , *Xiángmíng suàn fǎ*, 1373 AD) written by Hé Píngzǐ (何平子). These agree completely with the verses used nowadays.

6.4 The introduction of abacus calculation

The ultimate completion of the 'verses on division' (including 'testing divisor' verses and the 'starting from 1' verses) was a key step in the process of evolution from counting rod calculations to abacus calculations. Méi Wéndǐng (梅文鼎), a mathematician at the beginning of the Qīng Dynasty, said: 'The "verses on division" are very neat and clearly made, they foster the popularization of the abacus'. Méi Wéndǐng's remarks have an element of truth in them. After the final completion of the 'verses on division', in multiplications using counting rods each digit of the product was obtained by reciting a verse of the nine-nines rhyme; in counting rod division, each digit of the quotient can also be obtained by reciting a verse from the 'verses on division'. In this context it became more obvious that counting rod calculation as a means of calculating was not adequate for practical needs (it requires changing the counting configurations continually). Now on reciting '2 dividing 1, increase to 5' changing the starting place from 1 to 5 requires adding in four counting rods one by one. The movement of the hand is not as fast as recitation, and is even slower when compared with the speed of reasoning in calculation. Upon reaching this stage, counting rod calculation had arrived at the point where it had to change.

As a consequence, beads in an abacus replaced counting rods. Then, by stringing the beads together, just moving one of these beads in the abacus replaced putting in or taking out counting rods. At the same time the beads were differentiated into upper and lower compartments. In the lower compartment one bead stands for 1, in the upper compartment one bead stands for 5. This is similar to the rôle of the upper counting rod, which represents five in counting rod calculations. Consequently the abacus is modelled on counting rod calculations. It inherited certain rules from the methods of recording numbers using counting rods; so it has evolved from the foundations of the ancient counting rod calculations.

When was the abacus introduced? Who was its first inventor? These two questions remain without definite answers up to the present day. The situation just described shows that abacus calculation was not the invention of just one person but the product of an era; it was through the needs and demands of the populace in their everyday life that the abacus gradually developed and was finally completed.

After researching the relevant literature one can roughly infer that abacus calculation had been widely introduced into Chinese society at the very latest by the beginning of the 15th century.

As far as the earliest records of the abacus are concerned it is possible to trace these back to the end of the Yuán dynasty (the middle of the 14th century). In Táo Zóngyí's (陶宗儀) work *Talks while the Plough is Resting* (輟耕錄 , *Chuò gēng lù*) (1366 AD) there is a passage which uses three examples to illustrate the attitudes to work of maids and servants. He said:

'When maids and servants first arrive they are like 'loose beads on a board': you don't have to say anything, they work on their own. A little bit later they are like 'beads on a calculating board': you say something and they move. After a long time they are like 'beads on the head of Buddha': you talk to them all day, they stare at you and though you speak to them they still won't work.'

To the present day when Chinese are discussing someone with no initiative they still customarily use the same phrases. 'So and so is just like a bead on an abacus, you push him and then he moves.' What Táo Zóngyí is indicating is very much like the situation on an abacus. However in other places in *Talks while the Plough is Resting* there are reports and discussions of counting rod calculations, so it is worth asking whether what is called 'beads on the calculating board' really means 'beads' on the 'abacus' even though the characters used (算盤珠 , suàn pán zhū) nowadays do mean 'abacus beads'

The earliest book amongst those still in circulation to systematically introduce abacus calculation also contains drawings of the shape of an abacus and in addition verses on the methods for addition and subtraction. This is the *Method of Calculating on an Abacus* (盤珠算法 , *Pán zhū suàn fǎ*), produced in 1573 AD, the first year of the Wàn Lì reign (萬曆) of the Míng Dynasty, and collated by Xú Xīnlǔ (徐心魯). However, before that, there were some methods of abacus calculation recorded in Wú Jìng's (吳敬) book *Complete Description of the Nine Chapters on Arithmetical Techniques* (九章算法比類大全, *Jiǔ zhāng suàn fǎ bǐ lèi dà quán*, 1450 AD) and Wáng Wénsù's (王文素) book *Precious Mirror of Mathematics* (算學寶鑑, Suàn xué bǎo jiàn, 1524 AD), which could only be relevant to abacus calculations. These two texts contain the so-called 'starting from 5' (or 'building up from 5') verse and the 'going over 5' verse, both of which are obviously concerned with abacus calculations. The 'going over 5' verse says: 'Take away 1, remove 5 put down 4; Take away 2, remove 5 put down 3, . . .' This refers to subtracting 1 or 2 from a particular digit of a given number. In counting rod calculations the rule is '5 does not stand alone', i.e. 5 must be represented by ☰ or ⅢⅠⅠ , so there is no difficulty when subtracting 1 or 2 — this only requires removing one or two counting rods. It is only when wanting to subtract 1 in a place where there is no 1 [single bead] that it is required to 'take away 5 [represented by a single bead], put down 4 [below]' and if there are not two [single beads] when subtracting 2 that requires 'not 2, take away 5, put down 3 [below]'. The verses in these books look very much like the verses for abacus calculation. However, in their two books Wú Jìng and Wáng Wénsu did not explicitly mention abacus calculations.

Up to the end of the Ming Dynasty, besides Xú Xīnlǔ's *Method of Calculating on an Abacus* mentioned above, the extant works that mention abacus calculations refer to the following books: *Rules of Mathematics* (數學通軌, *Shù xué tōng guǐ*, 1578 AD) by Kē Shàngqián (柯尚遷), *A New Account of the Science of Calculation* (算學新說, *Suàn xué xīn shuō*, 1584 AD) by Zhū Zàiyù (朱載堉), the *Postscript to the Systematic Treatise on Arithmetic* (直指算法統宗, *Zhí zhǐ suàn fǎ tǒng zóng*, 1592 AD) and *Highlights of Calculation Methods* (算法纂要, *Suàn fǎ zuǎn yào*, 1598 AD) by Chéng Dàwèi (程大位), and *Directory of Calculation Methods* (算法指南, *Suàn fǎ zhǐ nán*, 1604 AD) by Huáng Lóngyín (黃龍吟). Among these the *Systematic Treatise on Arithmetic* by Chéng Dàwèi was the most widely circulated, it had the greatest influence, and its later editions continued to circulate. (See also Needham 1959 Vol. 3, pp. 74 ff.)

6.5 Chéng Dàwèi and his *Systematic Treatise on Arithmetic* (算法統宗, *Suànfǎ tǒngzóng*)

Chéng Dàwèi, (see Fig. 6.1) other name Rǔsī (汝思), alias Bīnqú (賓渠), was a native of Xiūníng (休寧) of Ānhuī (安徽) province. When he was 20 years old he was trading in the lower reaches of the Yangtze River and developed an interest in mathematics. He collected a lot of books and sought out expert teachers. After several decades of hard work, in his sixtieth year he

Fig. 6.1 Chéng Dàwèi.

finally wrote his book *Systematic Treatise on Arithmetic*, in 1592 AD. In his essay 'Postscript to the Systematic Treatise on Arithmetic' he recorded in detail the story of how he came to write the book:

I started to learn mathematics when I was young. About the age of 20 I was trading and travelling around the Wú (吳) and Chǔ (楚) districts and sought out expert teachers. Then I retired and studied very hard near the source of the river Shuài (率) for more than 20 years in order to collect together the fundamentals until at last I felt I had achieved something. So I put together the methods of the various schools and added my own ideas and having got that all together I wrote a book. Earlier methods which had not been fully worked out, I clarified; what was incomplete I supplemented; what was repetitive and superfluous I omitted; what was brief and fragmentary, I filled out with detail; and finally I rooted out the false and the nonsensical, put all in order and made the text lucid. . . .

Chéng Dàwèi expended a great deal of effort on this text.

The *Systematic Treatise on Arithmetic* is a practical mathematics text. It uses abacus calculation as the main means of computation, and the complete text contains 595 problems. Most of these problems were taken from other mathematical texts such as the *Methods of Calculation in the Nine Chapters* (九章通明算法 , *Jiŭ zhāng tōng míng suàn fă*, 1424 AD) by Liú Shilóng (劉仕隆) and the *Complete Description of the Nine Chapters on Arithmetical Techniques* (九章算法比類大全 , *Jiŭ zhāng suàn fă bĭ lèi dà quán*, 1450 AD) by Wú Jing. In the book *Systematic Treatise on Arithmetic* the chapters are arranged according to the layout of the *Nine Chapters on the Mathematical Art*; the complete book has 17 chapters.

The contents of Chapters 1 and 2 range from 'teachings of the earlier teachers' and 'outlines of methods of calculation' to items on 'large numbers' and 'small numbers' as well as descriptions of the systems for measuring length, weight and volume, field measurement, the methods for fixing the decimal point in abacus calculations, and verses for the methods of addition, subtraction, multiplication and division. These first two chapters clearly present the groundwork for understanding the rest of the book. The way of describing things is fairly similar to that in Zhū Shijié's book *Introduction to Mathematical Studies*. It is most important to note that the verses for the abacus calculations of addition, subtraction, and multiplication are essentially complete. Some of these verses are still in wide circulation today and have not been modified further.

Chapters 3–12 have the same titles as the chapters in the *Nine Chapters on the Mathematical Art*. The various sections in the chapters are as follows:

Chapter 3: 'Chapter 1, Field measurement.' At the very beginning of this chapter Chéng Dàwèi introduces a kind of instrument which he calls a 'walking measuring vehicle' (丈量步車 , Zhàng liàng bù chē), the equivalent of the present day tape measure.

Chapter 4: 'Chapter 2, Cereals and cloth' (the original name in the *Nine Chapters on the Mathematical Art* is 'Cereals').

Chapter 5: 'Chapter 3, Proportional distribution'.

Chapter 6 and 7: 'Chapter Four, "What width?"'. This chapter treats extracting square and cube roots, the corollary to extracting square roots, and the corollary to extracting cube roots (the methods of solution of quadratic and cubic equations). In this chapter he has copied the 'diagram on the source of opening the square' (Pascal's Triangle) from Wú Jing's book but nowhere in the book are there any problems recorded on extracting higher degree roots. On the 'source diagram' (Pascal's Triangle) Chéng Dàwèi wrote: 'Although this diagram was included in the *Nine Chapters* by Wú for extracting square roots up to sixth roots, he did not say how to use it and the explanation is incomprehensible.' One can see that at that time not only had the 'method of extracting roots by iterated multiplication' been lost but also no-one could really understand and use the method of extracting roots by means of the 'source diagram'.

Chapter 8: 'Chapter 5, Construction consultations'.

Chapter 9: 'Chapter 6, Fair taxes'.

Chapter 10: 'Chapter 7, Excess and insufficiency'. (In the *Nine Chapters on the Mathematical Art* it is called 'Excess and deficit'.)

Chapter 11: 'Chapter 8, Rectangular arrays'. In this chapter problems on systems of linear equations with two or three unknowns are called 'two colours', 'three colours', 'four colours' etc. When Chéng Dàwèi explains the meaning of 'rectangular array' he uses phrases such as 'required to arrange *n* columns', so it appears that the problems on systems of linear equations were not solved by using the abacus.

Chapter 12: 'Chapter 9, Gōugǔ'.

In Chapters 13–16 there are recorded several of what are called 'difficult problems'; these difficult problems were copied from the texts by Liú Shìlóng and Wú Jìng. Just as Chéng Dàwèi says, 'these "difficult problems", look difficult but in fact are not difficult, though the wording is tricky and ingenious so as to make the teachers of mathematics feel puzzled and not know how to start.'

Chapter 17 includes 26 types of 'miscellaneous methods'. Amongst these there is the 'method of multiplication by laying on the ground', which had been transmitted from the Arab countries in the west, also there is the method of 'a handful of gold' (a sort of quick calculation method not using the abacus) and various types of magic square, etc.

At the end of the *Systematic Treatise on Arithmetic* under the heading of 'The Sources of Mathematics' (算學源流 , Suànxué yuán lǐ) there is a list totalling 51 names of various mathematical texts from the Sòng and Yuán periods (960–1368 AD) and later. Of these only 15 have survived to the present. This bibliography is very useful for understanding the development of mathematics in the period from the Sòng and Yuán to the Ming.

The *Systematic Treatise on Arithmetic* is a very important work in the whole development of mathematics in ancient China. Throughout all the time it circulated there was no other mathematical work that could be compared to it for breadth or depth. In 1716 AD (the 55th year of the Kāng Xī (康熙) reign of the Qīng Dynasty), in the preface to the new reprint of the *Systematic Treatise on Arithmetic*, a descendant of the Chéng family wrote that since the appearance of the *Systematic Treatise on Arithmetic* in the Wàn Lì reign of the Ming in the year Rén Chén (壬辰 , 1592 AD), 'it has circulated widely in the country for more than a hundred years up to the present day. Mathematicians throughout the country all had a copy in their homes and in comparison this book can be counted as an authority in the same way as the Four Books and the Five Classics used by candidates for the civil service examinations'. This is by no means an exaggeration. (The Four Books and Five Classics form the canon of classical Chinese literature and occupy a place in Chinese culture akin to the Old Testament in Western cultures.)

The completion of the *Systematic Treatise on Arithmetic* and its wide distribution signal the completion of the evolution from counting rod to abacus calculation. From that time on the abacus became the main means of calculating, the counting rod calculations of ancient times were gradually forgotten and ultimately lost. Thereafter ordinary people only knew abacus calculations and did not know of the existence of counting rod calculations. This state of affairs continued until the middle of the 18th century when, as a consequence of the deep research of scholars of the Qing Dynasty into ancient mathematics, people began to understand the stages of the evolution from ancient counting rod calculations to abacus calculation.

7

The first entry of Western mathematics into China

7.1 A brief description of the situation when Western mathematics first entered

During the Wàn Lì (萬曆) reign of the Míng Dynasty (end of the 16th to beginning of the 17th century AD) there were big developments in the economy throughout China and in certain places some trades had begun to develop various capitalist methods of production. However because of extreme cruelty and extortion by the Míng rulers and also because of warfare, which continued for quite a number of years from the end of the Míng to the beginning of the Qīng period, the economy was unable to continue to expand and for a fairly lengthy period it was stagnant. On the other hand the Europe of those days was just the opposite. During the 15th and 16th centuries Europe gradually changed from a feudal to a capitalist society. By the 17th century this change was virtually complete.

As is well-known, the development of capitalism is inseparable from necessarily aggressive competition for raw materials, markets, and labour. In the 1580s countries in the West that had developed capitalism earlier started to penetrate the Far East and China. Besides the merchants who acted in piratical style, a lot of missionaries were also sent in the vanguard. From then on science from the West (including mathematics) gradually passed into China by way of the aggressive activities of the missionaries.

The missionaries who came to China in those days were mostly members of the Society of Jesus (Jesuits). The Society of Jesus is a conservative Catholic organization founded in the 16th century by southern European countries as a force for the conservative Roman Catholic religion, which was opposed to the Protestant Reformation. This organization was formed in 1540 AD. It established a theological institution in Rome and at the same time assigned its members who were trained there to various places throughout the world on missionary activities. The missionaries could use any sort of method to complete their missions of evangelization. For example, in Paraguay they used force to suppress the opposition of the natives and establish the despotic rule of the Church. At that time China was relatively strong under the rule of the Míng Dynasty, so the use of force was not practical. Science and technology therefore were used as a means of penetration.

In the year 1582 (the tenth year of the Wàn Lì reign of the Míng), the Jesuit priest Matteo Ricci (1552–1610 AD) arrived in Macao and a year later entered mainland China. He was the most important Jesuit priest to come to

China. When he arrived in Guǎngzhōu (廣州 , Canton) in order to buy their trust he gave to the local officials things he had brought along, such as a sundial, striking clocks, maps, a celestial sphere, etc. Thereafter he gradually moved inland to carry out his activities. In the 200 years from the end of the 16th to the 18th century several hundred missionaries besides Matteo Ricci came continually to operate in China. Most of them took Chinese names.

When Matteo Ricci arrived in China the Dà Tǒng calendar and the Islamic calendar were inaccurate: in particular, the most glaring errors concerned solar and lunar eclipses. In general the forecasts did not agree with actual events. At the same time knowledge about the manufacturing of cannon was required in order to defend the borders against the Manchu. The most urgent requirements of the Ming government at that time were the emendation of the calendar and the manufacture and use of cannon. At that time too some educated patriots were very anxious to strengthen the country both economically and militarily and consequently showed intense interest in the science and technology from the West. The missionaries used this point first of all and started on the amendment and production of the calendar through this connexion with the educated circles of the upper class. They obtained the Emperor's confidence as a means of achieving their aims of evangelization and other aggressive activities.

The first phase of Western mathematics passing into China centred on the emendation of the calendar. This phase lasted from Matteo Ricci's arrival in China (1582 AD) until the Yōng Zhēng (雍正) reign of the Qīng (that is, the beginning of the 18th century): in all more than 150 years. The main historical events during this phase of the entry of Western mathematics into the East were:

(a) in the initial stages translation into Chinese of two books: Euclid's *Elements of Geometry* and Clavius's *Epitome of Practical Arithmetic*;

(b) during the end of Ming and the beginning of the Qīng availability of the mathematical knowledge for the computation of the Western calendar;

(c) the mathematical research of Méi Wéndìng (梅文鼎 , 1633–1721 AD); and

(d) The editing of the *Collected Basic Principles of Mathematics* (數理精蘊 , *Shù li jīng yùn*) under Emperor Kāng Xī (康熙).

In the initial phase the most important elements of the mathematics entering China were Euclid's geometry, the techniques of calculating with pen and paper, (plane and spherical) trigonometry, and logarithms.

7.2 The translation of the *Elements of Geometry* and the *Epitome of Practical Arithmetic*

1 Euclid's *Elements* and Xú Guāngqǐ (徐光啓)

Euclid's *Elements* and the *Epitome of Practical Arithmetic* were the first Western works translated into Chinese. These two books were brought in by the first missionary in China, Matteo Ricci. The *Elements of Geometry* 'were orally translated by Matteo Ricci, recorded by Xú Guāngqǐ of Wú Sòng (吴淞)' and the *Epitome of Practical Arithmetic* 'dictated by Matteo Ricci of the Western Sea, edited by Lǐ Zhīzǎo (李之藻) of Zhè Xī (浙西)'. As mentioned above, when Western science and technology first entered China some educated patriots who were anxious to strengthen the country economically and militarily showed intense interest in the knowledge of Western science and technology. Xú Guāngqǐ and Lǐ Zhīzǎo are representative of these people.

Let us turn first to Xǔ Guāngqǐ (Fig. 7.1) and the translation of Euclid's *Elements*.

Xǔ Guāngqǐ (徐光啓 , 1562–1633 AD), other familiar name Zǐxiān (子先), also known as Xuánhù (玄扈), was a native of Wú Sòng (吴淞) (in the present-day city of Shànghǎi). For some time he held the office of Grand Secretary of the Wén Yuān Institute (文淵閣 , Wén yuān gé) and so was 'the first man in China after the monarch himself' (see Väth 1933, p. 103, quoted in Dunne 1962, p. 220) from the end of the Wàn Lǐ reign through the Tiān Qǐ (天啓) and Chóng Zhēn (崇禎) reigns. He was well versed in astro-

Fig. 7.1 Xú Guāngqǐ (Paul Xú).

nomy and the computation of the calendar and was the principal figure in the reform of the calendar at the end of the Míng Dynasty. He also did some research in agriculture, commented on various earlier agricultural texts, and wrote the famous book entitled *Complete Treatise on Agriculture* (農政全書 , *Nóng zhèng quán shū*), which has altogether 60 chapters with a total of more than 600 000 characters. Towards the end of the Míng Dynasty the Manchu rulers started attacking along the northern borders and Xú Guāngqǐ memorialized the Emperor several times on military affairs. At the same time he was training the army in new techniques in Tōng Zhōu (通州) and his opinion favoured adopting the Western arsenal, in particular Portuguese cannon from Macao. He was a scientist who loved his country.

Before he became an official in the capital Xú Guāngqǐ taught in Shànghǎi, Guǎngdōng (廣東) and Guǎnxi (廣西) provinces. During this period he read widely. While he was in Guǎngdōng he made initial contacts with some Western missionaries and began to get acquainted with the Western civilization brought in by them. In 1600 he met Matteo Ricci in Nánjīng and after that over a long period they saw each other frequently in Běijīng.

Together he and Matteo Ricci translated the *Elements of Geometry*. In 1607 they completed the translation of the first six books.

The *Elements of Geometry*, as everyone knows, is a piece of outstanding ancient Greek mathematics. It was compiled by the ancient Greek mathematician Euclid (about the fourth century BC), so it is often called Euclid's *Elements*. The whole work is in 13 books: Books 1 to 6 are on plane geometry, Books 7 to 10 are on number theory and Books 11 to 13 are on solid geometry. Later on two extra books were added to the original 13 books. Of these Book 14 is an addendum to Book 13, and Book 15 was added in the third century AD; it is of much less value.

The *Elements of Geometry* was preserved in Arabic from the Middle Ages. In the 10th century AD it was translated from Arabic into Latin and passed into various countries in Western Europe. To the present day, the basic content of the *Elements* is one of the fundamental subjects still taught in secondary schools throughout the world. In general, textbooks on plane and solid geometry still follow the system used in the *Elements of Geometry* in their essentials.

The translation of Xú Guāngqǐ and Matteo Ricci was based on Clavius's *Euclidis elementorum libri XV*. Their copy of Clavius is now in the Běijīng library, where it was taken from the North Hall (北堂 , Běi táng) library. Clavius was a German who was a professor in the Jesuit College in Rome where Jesuits were trained. He was Matteo Ricci's teacher and also a friend of both Kepler and Galileo (see Dunne 1962, p. 24). His works were written in Latin. The word 'clavius' in Latin means 'nail' (釘 , dīng)', so Matteo Ricci and Xú Guāngqǐ often referred to Clavius as Mr. Dīng (丁先生, Dīng xiānshēng) because the character 丁 (Dīng) is pronounced almost the same as the character for 'nail' (釘). Clavius was well versed in both mathematics and astronomy. Many of the various books on mathematics and astronomy brought into China by Matteo Ricci and other missionaries were written by Clavius. (See the catalogue of the North Hall library.)

The translation of Xú Guāngqǐ and Matteo Ricci was of just the first six books of the *Elements of Geometry*. At that point Xú Guāngqǐ wanted to translate the whole book but Matteo Ricci did not wish to do so. In the preface to the translation of the *Elements of Geometry* Matteo Ricci wrote:

The Grand Scholar [that is, Xú Guāngqǐ] is very enthusiastic, he wanted to complete the translation but I said: No, please first distribute it. Help those interested to study it and if it proves useful, then we shall translate the rest. The Grand Scholar said: Very well. If this book is useful it should be used. It does not have to be completed by us. So we shall stop the translation and publish it . . .

However in a later edition of the translation of the *Elements of Geometry*, Xú Guāngqǐ said: '. . . we do not know when and by whom this important task will be continued.' He still felt very unhappy about this. [The translation was finally completed by A. Wylie and Lǐ Shànlán (李善蘭) in 1857 AD.]

The *Elements of Geometry* was the first work translated from Latin in China. At the time of the translation there was nothing like a technical dictionary to follow so a lot of the translations of terms had to be invented for the occasion. There is no doubt this required careful research and a great deal of mental effort. Many of the translations of technical terms are very appropriate and are not only still used in China up to the present day but have also influenced countries such as Japan and Korea. For example, the terms 'point' (點 , diǎn) 'line', (綫 , xiàn), 'straight line', (直綫 , zhí xiàn), 'curve' (曲綫 , qū xiàn), 'parallel lines' (平行綫 , píng háng xiàn), 'angle' (角 , jiǎo), 'right angle' (直角 , zhí jiǎo), 'acute angle' (銳角 , rui jiǎo), 'obtuse angle' (鈍角 , dùn jiǎo), 'triangle' (三角形 , sān jiǎo xíng), and 'quadrilateral' (四邊形 , sì biān xíng) were all first determined by this translation. Among the terms only a few have been altered later such as 'equilateral triangle' (等邊三角形 , děng biān sān jiǎo xíng), which at that time Xú Guāngqǐ translated as 'even side triangle' (平邊三角形 , píng biān sān jiǎo xíng), 'relative to' (比 , bǐ) which Xú Guāngqǐ translated as 'ratio' (比例 , bǐ lì) while ratio, (比例 , bǐ lì) was translated as 'rational ratio' (有理的比例 , yǒu lǐde bǐ lì) — and so on.

The *Elements of Geometry* is logically precise and its mode of presentation is completely different from the traditional Chinese *Nine Chapters on the Mathematical Art*. The *Elements of Geometry* starts with a few postulates and common notions and proceeds in a deductive manner, while the *Nine Chapters on the Mathematical Art* gives three to five examples then the general method of solution is presented: it adopts the inductive form. Xú Guāngqǐ had a relatively clear understanding of this special feature of the *Elements of Geometry*, which is quite different from the traditional mathematics of China. In the preface to the translation of the *Elements of Geometry* he says:

[The Elements] proceeds from the evident to the particular details, from doubt to certainty. What appears useless is very useful, in fact it is the foundation of everything [postulates and common notions appear useless but in fact they are the 'foundation of everything'], it is true to say that it is the basic form of the myriad forms, the medium for a hundred schools of learning.

In the essay 'Discourse on the *Elements of Geometry*', in particular, it further suggests that

This book has four 'no needs': no need to doubt, no need to guess, no need to test, no need to change. It has four 'cannots': cannot elude it; cannot argue against it; cannot simplify it; cannot try to change its order. There are three 'supremes' and three 'cans': it looks unclear, in fact it is supremely clear, so we can use its clarity to understand other unclear things; it looks complicated, in fact it is supremely simple, so we can use its simplicity to simplify the complexity of other things; it looks difficult, in fact it is supremely easy, so we can use its ease to ease other difficulties. Ease arises from simplicity and simplicity from clarity; finally, its ingenuity lies in its clarity.

He concludes that the whole book's ingenuity can be described in one word, 'clarity' (明 , Míng): this is to point out the special characteristic of logical deduction. So he maintained that 'For geometry considered as a branch of learning, if you understand it, you understand it thoroughly, if you find it obscure, everything is obscure.' Again, he says: 'The advantage of this book is that it can make its student humble himself and train his concentration; its student can apply the methods and increase his mental power; so no-one in the world should fail to study it.' However he knew that there were not many people at that time who would understand it so he continued: 'I predict that after a hundred years everyone will study it.' He fully understood that the ideas of geometry develop the ability to think logically and train people to become expert in problems of finding areas and volumes.

Although matters did not reach the stage of 'everyone study[ing] it', after its publication the *Elements of Geometry* nevertheless had a definite influence from the point of view of later mathematics. The *Elements* and the *Epitome of Practical Arithmetic* became the mathematical groundwork for learning the Western methods for computing the calendar. In deciding to translate these two technical books, part of the missionaries' aim was to motivate the emendation of the calendar.

During the Kāng Xī (康熙) reign of the Qīng an encyclopaedia of mathematics entitled the *Collected Basic Principles of Mathematics* (數理精蘊 , *shù lǐ jīng yùn*, 1723 AD) was compiled, and it contained the *Elements of Geometry*. However this was translated from an 18th century French geometry textbook and there are a lot of differences from Euclid's *Elements*.

The translation of the last nine books of the *Elements of Geometry* was completed by Lǐ Shànlán (李善蘭 , 1811–1882 AD) and Alexander Wylie in 1857, towards the end of the Qīng period. The civil service examinations were abolished towards the end of the Qīng Dynasty and Western-style schools were established. By then geometry had become a compulsory subject in schools. By this time the situation predicted by Xú Guāngqi that 'everyone will study it, [Euclid's *Elements*]' was fulfilled. However, at this time the geometry textbook was vastly different in form from the *Elements of Geometry*.

In the period from the end of Míng to the beginning of Qíng, besides the joint translation of the first six books of the *Elements of Geometry* by Xú Guāngqi and Matteo Ricci, in the *Chóng Zhēn Reign Treatise on (Astronomy and) Calendrical Science* (崇禎曆書 , *Chóng zhēn lì shū*), edited during the Chóng Zhēn reign toward the end of Míng, there is also a text entitled the *Essentials of Geometry* (幾何要法 , *Jǐ hé yào fǎ*) by Giulio Aleni (1582–1649). Aleni was a Jesuit cartographer who followed Ricci to China. In the book there are various problems on the construction of plane figures. The arrangement of the book is very different from Euclid's *Elements*. In particular the problem of the quadrature of the circle is not in Euclid's *Elements*. Further, the *Chóng Zhēn Reign Treatise on (Astronomy and) Calendrical Science* also includes such texts as the *Complete Theory of Surveying* (測量全義 , *Cè liáng quán yi*, 1631 AD) and the *Complete Surveying* (大測 , *Dà cè*, 1631 AD). This book frequently used definitions that are outside the scope of the first six books of Euclid but which occur in Book 9 and Book 13. This is a brief description of the situation during the initial stages when geometry entered China.

2   Lǐ Zhīzǎo (李之藻) and the *Epitome of Practical Arithmetic*

The other mathematical text translated into Chinese about the same time as the *Elements of Geometry* was the *Epitome of Practical Arithmetic* (1631 AD) known in Chinese as 同文算指 (*Tóng wén suàn zhǐ*, the *Treatise on European Arithmetic*). The complete work, *Epitome of Practical Arithmetic*, has a total of eleven chapters and was translated jointly by Matteo Ricci and Lǐ Zhīzǎo (Fig. 7.2).

Lǐ Zhīzǎo (李之藻 , 1565–1630 AD), other name Zhènzhī (振之 ; also Wǒcún, 我存), also known as Liáng Ān (涼庵), was a native of Rén Hé (仁和 ; the present-day city of Hángzhōu). He met Matteo Ricci in Nánjīng and was a friend of Xú Guāngqi. Later he was recommended by Xú Guāngqi for participation in the work of emending the calendar at the end of the Míng Dynasty.

The *Treatise on European Arithmetic* (同文算指 , *Tóng wén suàn zhǐ*, literally the *Combined Languages Mathematical Indicator*) introduces the arithmetical knowledge of Western mathematics. This book is principally based on the translation of Clavius's *Epitome of Practical Arithmetic*. (The Latin original, which was in the North Hall Library, is now in the Běijīng Library.) The book also treats some methods of calculation with reference to the traditional Chinese mathematical texts which were not known in the

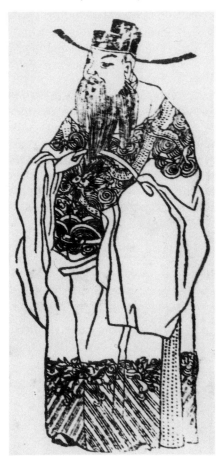

Fig. 7.2 Lǐ Zhīzǎo.

West. The complete work is divided into three volumes, *Introduction, General Survey*, and *Supplement*. In Lǐ Zhīzǎo's preface it says: '*Introduction*: a summary of the principal ideas, about half [the whole work]. *General Survey*: brief exposition of examples which are easy to read, including work from the *Nine Chapters* which is not outside the scope of the original [Clavius's] text. *Supplement*: a survey of various techniques of measuring circles so as to preserve them for later readers.' The *Introduction* presents the methods for locating the decimal point in calculations and the methods for the four operations of addition, subtraction, multiplication, and division in pen and paper computations. The *General Survey* is the core of the whole work with a total of eight chapters and 18 sections. It describes the methods of calculation for fractions, ratios, sums of series, 'excess and deficit', 'rectangular arrays', 'extracting roots', 'corollary to extracting roots', etc. Some people believe that in some of the content of this work Lǐ Zhǐzǎo was 'using Western methods to rewrite the *Nine Chapters*', but this is not correct. Let us compare the *Introduction* and the *General Survey* in the *Treatise on European Arithmetic* with Clavius's original text. Then it will be found that the order in the *Introduction* is the same as in Clavius's book and the order in the *General Survey* volume is also the same; what is different is that Lǐ Zhǐzǎo, following Chinese mathematical tradition, has added a lot of problems and methods of calculation. The *Supplement* was not printed so it has not been passed down to the present.

The most important part of this book *Treatise on European Arithmetic* is the systematic introduction of Western methods of calculation using pen and paper. In the section 'Fixing the position' in Chapter 1 of the *Introduction*, Lǐ Zhǐzǎo first introduces the method of writing numbers used in the West at that time. He says:

In using pen and paper to represent abacus calculations start with 1 up to 9 and according to what the number is write it down on the paper. Getting to ten do not write down ten (十 , shi) but write down 1 advanced one position to the left and put 0 in the original position . . . from ten advance to a hundred and from a hundred to a thousand, from a thousand to ten thousand and so on.

Calculation using paper and pen is a completely different technique from counting rod or abacus calculation. No matter how the numbers are recorded

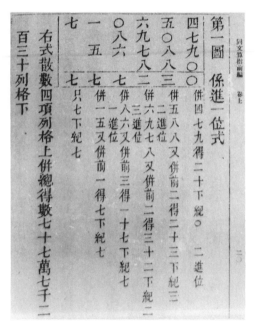

Fig. 7.3 The method of addition in the *Treatise on European Arithmetic*.

or what the method of calculation there is always some difference. In the *Treatise on European Arithmetic* the Arabic numerals, 1, 2, 3, . . ., 0 have not been adopted and the usual Chinese characters 一 , 二 , 三 , . . ., 〇 are used to record numbers and perform calculations.

Next in the *Treatise on European Arithmetic* the methods for addition and subtraction with pen and paper are introduced. This is exactly the same method as the ordinary calculations for addition and subtraction with pen and paper we use today. They all start calculating from the last digit, that is from right to left; calculation proceeds progressively from the smallest place to the biggest. Figure 7.3 is an example of addition, namely 710654 + 8907 + 56789 + 880 = 777230. The only difference from the method of addition and subtraction used today is that the numerals are different.

Multiplication also starts from the last digit, from right to left, and proceeds progressively from the smallest place to the biggest. That is, the calculation starts with the smallest place in the multiplier. This is just the opposite of the traditional method of multiplication in China using counting rods, which starts from the largest and goes to the smallest place, that is, starts the calculation from the highest place. The method of multiplying 394 × 38 = 1497 is shown in Fig. 7.4.

The method of division introduced in the *Treatise on European Arithmetic* is the so-called 'Galley Method' that we have already described in Chapter 1

Fig. 7.4 The method of multiplication in the *Treatise on European Arithmetic*.

of this book when we were introducing methods for division using counting rods. The 'Galley Method' is not quite the same as the method of using pen and paper commonly employed nowadays. It was not until the time of the book the *Collected Basic Principles of Mathematics* (1723 AD), compiled during the Kāng Xī reign of Qīng, that the present day form of the method of division was introduced.

At the same time as the methods for addition, subtraction, multiplication, and division using pen and paper were introduced in the *Treatise on European Arithmetic* the method of 'checking calculations' was also brought in. This is the way of checking whether the result is accurate after making the calculation. This is just the part missing in the traditional mathematics of ancient China. Besides 'using subtraction to check addition', 'using addition to check subtraction', 'using division to check multiplication' and 'using multiplication to check division' there are also the so-called methods of 'subtracting nines' and 'subtracting sevens' for testing calculations. The method of 'subtracting nines' is also known in the West as the 'method of casting out nines'. The 'method of casting out nines' can first be found in a 10th century Hindu mathematics book. It was first introduced in China in the *Treatise on European Arithmetic*.

For calculating with fractions, in the *Treatise on European Arithmetic* the denominator is put on top with the numerator at the bottom. For example, $\frac{4}{7}$, $\frac{3}{5}$, and $\frac{25}{48}$ are respectively recorded in the *Treatise on European Arithmetic* as

七	五	四八
四	三	二五

that is:

7	5	48
4	3	25

This sort of method of recording fractions was not only different from the traditional practice of ancient China, it was also different from the method used in the West at that time. This way of recording fractions gives no advantage in convenience and it also made the Chinese people reading it feel uneasy with it. However this type of method of recording fractions continued till toward the end of the Qīng Dynasty, when it was corrected. When reading mathematical texts of this period one must be careful, otherwise one could take $\frac{4}{7}$ as $\frac{7}{4}$.

The various methods of calculation with fractions and ratios in the *Treatise on European Arithmetic* which came into China did not go beyond the limits of the achievements in the traditional mathematics of ancient China. One can say there was nothing new.

On the other hand Lǐ Zhīzǎo included a lot of methods of calculation from the ancient mathematical texts in China that were not present in Western mathematics. For example, in the book there are the method of solution of systems of linear equations — the 'method of rectangular arrays' (including the techniques of positive and negative), the method for finding a root of a quadratic equation — the 'method of the corollary on extracting roots', and the method to extracting higher order roots (Lǐ Zhīzǎo included an example of extracting an eighth root). These were not among the Western methods of calculation.

Arithmetic series (in the book called the method of successive addition), geometric series (method of iterated addition) and the formulae for finding the sums are also presented in the *Treatise on European Arithmetic*. However these were already to be found in the traditional mathematics of China at an early stage.

In conclusion it is possible to see that besides the techniques of pen and paper calculations there are a lot of other methods of calculation in the *Treatise on European Arithmetic* that had existed in the traditional mathematics of ancient China at an early stage. Lǐ Zhīzǎo used the Chinese calculations to augment the Western calculations in some parts and this indicates the areas where the Chinese calculations went beyond the Western calculations. From this we can see that although during the several centuries of the Mǐng Dynasty in ancient China mathematics appeared to be stagnating in some areas, nevertheless in other areas, in particular in algebra, it was still clearly in advance of the Western mathematics introduced by Matteo Ricci.

7.3 The emendation of calendars and mathematical knowledge in the various books on the calendar

1 A brief description of emending the calendar during the end of the Mǐng and the beginning of the Qīng Dynasty

As mentioned in the previous section, the initial phase of the entry of Western mathematics into China was basically oriented towards emending the calendar. Apart from the books on Catholicism, most of the texts were on Western astronomy, the computation of the calendar, and mathematics. Some Western mathematical knowledge came in through the works on the computation of the calendar. Below, we first describe the general situation regarding the emendation of the calendar at that time and then discuss the mathematical knowledge that entered by way of the various books on the calendar.

The calendar used in the Ming Dynasty was called the *Dà Tǒng Calendar* (大統曆 , Dà tǒng lì). The data and the methods of calculation used in the *Dà*

Tǒng Calendar were directly inherited from the *Works and Days Calendar* (1280 AD) written by Guō Shǒujīng and co-workers. No matter how accurate the *Works and Days Calendar* might have been when it was first compiled, because of the long period of use without changes, there was no way that the errors which occurred could have been avoided. So near the end of the Míng Dynasty a lot of the forecasts from the *Dà Tǒng Calendar* disagreed with observation; in particular the errors in forecasting solar and lunar eclipses were very obvious. Thus the emendation of the computation of the calendar became an important task at the end of Míng. At that time there were not many people who understood astronomy and the computation of the calendar very well; most of the officials in the State Observatory were conservative and lacked the ability to emend the computation of the calendar.

During the Míng Dynasty, in addition to the *Dà Tǒng Calendar* there was the so-called *Islamic Calendar* (回回曆 , Hui hui lì). It was compiled by the Islamic Department of the State Observatory and was the calendar for the Islamic people. The *Islamic Calendar* was based on the method of computing the calendar that had come into China from the Arab countries in the 13th century AD. By this time it had significant errors.

At this time the missionaries in China clearly perceived this point and so they got themselves acquainted with the educated upper class through work on emending the computation of the calendar. Their ultimate intention was to gain the confidence of the Emperor. In this way they used the suggestions for emending the calendar as the key to evangelizing and other sorts of activity in China. Matteo Ricci knew the limitations of his knowledge so he wrote letters requesting more missionaries well-versed in computing the calendar to be sent to China. Among the missionaries who came to China during this period, those well known in the field of calendar computation were:

Niccolo Longobardi (1559–1654, Italian, arrived in China in 1597);

Johann Terrenz Schreck (1576–1630, Swiss, arrived in China in 1621);

Johann Adam Schall von Bell (1591–1666, German, arrived in China in 1622);

Giacomo Rho (1593–1638, Italian, arrived in China in 1624).

Those who came to China later in the beginning of the Qīng Dynasty were:

Jean Nicolas Smogulecki (1611–1656, Polish, arrived in China in 1646);

Ferdinand Verbiest (1623–1688), Belgian, arrived in China in 1659).

The earliest proposal to emend the calendar was made in the Wàn Lì (萬曆) reign of the Míng Dynasty. According to the report in the *History of the Ming Dynasty* (明史 , *Ming shi*): 'On the first day of the eleventh moon in the year Gēng xū (庚戌) of the Wàn Lì reign [1610 AD] a solar eclipse occurred. The forecast of the Calendar Official was wrong. The Government proposed emending the calendar.' At that time some officials supported adopting the Western methods for computing the calendar, others were against the proposal, so from time to time there were discussions about emending the calendar. The ultimate move to reform the computation of the calendar came during the Chóng Zhēn (崇禎) reign. It is recorded that:

A solar eclipse occurred on the first day of the fifth moon in the second year of the Chóng Zhēn (崇禎) reign [1629 AD]. On the twenty-ninth day of the fourth moon the Ministry of Rites announced the forecasts of the solar eclipse by the three schools. The three schools were: *Dà Tŏng Calendar, Islamic Calendar* and the 'New Method' [that is the forecast of Xú Guāngqi according to Western methods of computing the calendar]. On the day of the eclipse Xú Guāngqi's forecast agreed with observation. On the fourteenth day of the seventh moon [of the same year] Xú Guāngqi was appointed to direct the emendation of the calendar and at the same time Li Zhizǎo (李之藻) was appointed. Xú [Guāngqi] recommended Nicolaus Longobardi, Johann Terrenz, Johann Adam Schall von Bell and Giacomo Rho as assistants for emending the calendar in the Department of the Calendar. Xú Guāngqi's policy was to 'gradually emend the calendar successively' and to 'seek out the source [of the Western Method]' so his opinion was: 'To improve [on the Western Method] demands thorough understanding, thorough understanding first requires translation.'

From the fourth year of the Chóng Zhēn reign (1631 AD) to the seventh year of the Chóng Zhēn reign (1634 AD) there were altogether 137 books of calendric texts presented to the Emperor on five occasions. These were the famous *Chóng Zhēn Reign Treatise on (Astronomy and) Calendrical Science* (崇禎曆書 , *Chóng zhēn lì shū*). Xú Guāngqi died in 1633 AD. Before he died he recommended Li Tiānjīng (李天經 , 1579–1659 AD) to continue the work of emending the calendar. The fourth and fifth submissions of the calendar were made by Li Tiānjīng. This massive work involved the Jesuits Johann Adam Schall von Bell, Johann Terrenz Schreck, Giacomo Rho, and Niccolo Longobardi as well as the Chinese Xú Guāngqi, Li Zhizǎo, and Li Tiānjīng. (See Needham 1959, Vol. 3, p. 447.)

Book thirty one of the *History of the Ming Dynasty* (明史 , *Ming shi*) says:

Seventh year [of the Chóng Zhēn reign, 1634 AD], Wèi Wénkui (魏文魁) reported: the forecasts of eclipses, solstices and equinoxes by the Official for Computing the Calendar were in error. Consequently Wénkui was ordered to the capital for an investigation. At that time there were four schools computing the calendar: Besides the Dà Tŏng and the Islamic, the Western Branch using the Western method and the Eastern Branch headed by Wénkui had been founded. Different schools had different opinions, their differences had to be settled in public debate.

These four schools argued against each other. After testing the various theories on several occasions and in particular their forecasting of solar and lunar eclipses, it was established by observation that the Western method was the most accurate. In the eighth moon of the 16th year of the Chóng Zhēn reign (1643 AD) the Ming government decreed that the court had decided to change the calendar and that the Western method would be adopted. However, shortly afterwards the Manchu army penetrated the Great Wall and the Míng Dynasty ceased to exist.

After the Manchu army penetrated the Great Wall the missionaries were relied on even more for the task of editing and computing the calendar. The Qīng government of those days put Johann Adam Schall von Bell in charge of the seal of the State Observatory in the second year of the Zhēn zhì (順治) reign, 1645 AD. From the second year of the Zhēn zhì reign the government decreed the adoption of the new calendar based on the Western calendar, which was called the Shíxiàn Calendar (時憲曆 , Shí xiàn lì). In that year Johann Adam Schall von Bell amended the *Chóng Zhēn reign Treatise on (Astronomy and) Calendrical Science* and developed it into the *Treatise on (Astronomy and) Calendrical Science according to the New Western Methods* (西洋新法曆書 , Xī yáng xīn fǎ lì shū) (13 volumes, a total of 100 books), and presented it to the Emperor.

These two sets of books, the *Chóng Zhēn reign Treatise on (Astronomy and) Calendrical Science* and the *Treatise on (Astronomy and) Calendrical Science according to the New Western Methods* brought together in printed form the contents of the books on astronomy, the computation of the calendar, and mathematics that had been brought into China. They have become extremely valuable material for understanding the history of this period, when Western astronomy and calendrical computation entered China. Of the collection of works the important ones as far as mathematical knowledge is concerned are the following:

Giulio Aleni: *Essentials of Geometry* (幾何要法 , Jǐ hé yào fǎ) four chapters;

Johann Terrenz: *Complete Surveying* (大測 , Dà cè), two chapters, *Tables of trigonometric functions* (割圓八綫表 , Gē tú bā xiàn biǎo) six chapters, *Brief Description of the Measurement of the Heavens* (測天約說 , Cè tiān yuē shuō) two chapters;

Johann Adam Schall von Bell: *Theory of the Celestial Sphere* (渾天儀說 , Dōng tiān yí shuō) five chapters, *Joint Translation of Tables and Diagrams of Trigonometric Functions* (共譯各圖八綫表 , Gòng yì míng tú bā xiàn biǎo) six chapters;

Giacomo Rho: *Complete Theory of Surveying* (測量全義 , Cè liáng quán yì) ten chapters, *Manual for proportional dividers* (比例規解 , Bǐ lì guī jiě) one chapter, *Calculation with rods* [Napier's bones] (籌算 , Chōu suàn) one chapter.

As well as the missionaries who were appointed to the State observatory to participate in correcting the calendar there were also other missionaries, not in the observatory, who introduced scientific knowledge to China. While the Polish missionary Nicolas Smogulecki was evangelizing in Nánjīng, Xuē Fèngzuò (薛鳳祚 , ?–1680 AD) and Fāng Zhōngtōng (方中通 , 1633–1698 AD) studied under him. Xuē Fèngzuò also published a collection of books called *Understanding Calendar Making* (曆學會通 , Lì xué huì *tōng*) jointly with him. (There were prefaces written in 1652 and 1654.) Among these the most important were two volumes introducing logarithms: *New Tables for Four Logarithmic Trigonometric Functions* (比例四綫新表 , *Bǐ lì sì xiàn xīn biǎo*) one chapter, and *Logarithm Tables with Explanations* (比例對數表 , *Bǐ lì duì shù biǎo*) one chapter.

Of the mathematics introduced in the works on the calendar that in the *Chóng Zhēn resign Treatise on (Astronomy and) Calendrical Science* and the *Treatise on (Astronomy and) Calendrical Science according to the New Western Methods* is of one sort, and that in *Understanding Calendar Making* is of another. The substance of these two sorts of mathematics will be discussed in detail in the following sections.

From the moment Western methods of calendrical computation entered China there were frequent clashes over the old and new methods. This sort of argument was in many cases tied up with anti-foreign feelings. Amongst them the most notable was the 'Yáng Guāngxiān (楊光先) case' at the beginning of the Kāng Xī (康熙) reign of Qīng (1664 AD) when Yáng Guāngxiān claimed, 'It is better to have no decent calendar than have Westerners in China.' However, when all the missionaries in China at that time had been confined and the Courts had appointed Yáng Guāngxiān to make the amendments to the calendar, his inadequate knowledge of calendar making resulted in frequent errors in forecasts, and finally the Western missionaries were reinstated in the observatory (see Dunne 1962, p. 360).

Yáng Guāngxiān's defeat finally led to the firm establishment of the Western calendar methods. From then on the State Observatory was controlled by foreign missionaries for about 200 years.

2 Mathematics in the various calendar books

There is a lot of mathematics included in the collections *Chóng Zhēn reign Treatise on (Astronomy and) Calendrical Science, Treatise on (Astronomy and) Calendrical Science according to the New Western Methods* and *Understanding Calendar Making*. Amongst this the most important items are plane and spherical trigonometry and the tables that were required for such mathematics. In addition Napier's bones, Galilean dividers and several other calculation devices were introduced in these books. These Western methods of calculation were brought into China at that time.

The methods of plane and spherical trigonometry

The *Chóng Zhēn reign Treatise on (Astronomy and) Calendrical Science* contains the *Complete Surveying, Tables of Trigonometric Functions* (割圓八綫表 , *Gē tú bā xiàn biǎo*) and the *Complete Theory of Surveying* (測量全義 , *Cè liáng quán yì*), all of which were published in 1631 AD. All these texts contain material on plane and spherical trigonometry. Also, in the collection of books *True Course of Celestial Motions* (天步眞原 , *Tiān bù zhēn yuán*) written by Nicolas Smogulęcki and Xuē Fèngzuò, there is a book

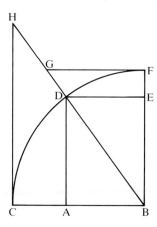

Fig. 7.5 The 'eight lines' diagram.

entitled *Method of Trigonometrical Calculations* (三角算法 , *Sān jiǎo suàn fǎ*, 1653 AD).

In all of these books the lengths of line segments were used to define the meanings of trigonometrical functions. For example in Chapter 7 of the *Complete Theory of Surveying* it says: 'Each arc and each angle has eight types of lines: sine (正弦 , zhēng xiàn), tangent (正切綫 , zhēng qiē xiàn), secant (正割綫 , zhēng gē xiàn), versine (正矢 , zhēng shǐ), cosine (餘弦 , yú xián), cotangent (餘切綫 , yú qiē xiàn), cosecant (餘割綫 , yú gē xiàn), and coversine (餘矢 , yú shǐ). Using B as centre, as in Fig. 7.5, describe a circle with radius BC. Then the trigonometric functions are defined for the angle CBD or the arc CD by the lengths of the various line segments — the 'eight lines'. Here AD is the sine, CH is the tangent, BH the secant, AC the versine, DE the cosine, GF the cotangent, BG the cosecant, and GF the coversine. (The Chinese characters mean 'cutting' in various senses.)

There were relatively detailed records of various formulae of plane trigonometry in the *Complete Theory of Surveying*. The following formulae (apart from the explanations in brackets) are all from that book. At that time in Europe too there was no notation for trigonometric functions; the formulae were written out in words. Using modern notation these formulae were equivalent to:

$$
\begin{cases}
\sin \alpha \cdot \operatorname{cosec} \alpha = 1 \\
\cos \alpha \cdot \sec \alpha = 1 \\
\tan \alpha \cdot \cot \alpha = 1 \\
\tan \alpha = \dfrac{\sin \alpha}{\cos \alpha} \\
\cot \alpha = \dfrac{\cos \alpha}{\sin \alpha} \\
\sin^2 \alpha + \cos^2 \alpha = 1
\end{cases}
$$

$$\frac{a}{\sin A} = \frac{b}{\sin B} = \frac{c}{\sin C}$$

$$\begin{cases} c^2 = a^2 + b^2 - 2ab \cos C \\ b^2 = c^2 + a^2 - 2ac \cos B \\ a^2 = b^2 + c^2 - 2bc \cos A \end{cases}$$

$$\tan \frac{A - B}{2} = \frac{a - b}{a + b} \tan \frac{A + B}{2}$$

$$\begin{cases} sin(\alpha \pm \beta) = \sin \alpha \cos \beta \pm \cos \alpha \sin \beta \\ \cos(\alpha \pm \beta) = \cos \alpha \cos \beta \mp \sin \alpha \sin \beta \end{cases}$$

$$\begin{cases} \sin 2\alpha = 2 \sin \alpha \cos \alpha \\ \sin \frac{\alpha}{2} = \sqrt{\left[\frac{1 - \cos \alpha}{2}\right]} \end{cases}$$

$$\sin \alpha = \sin(60° + \alpha) + \sin(60° - \alpha)$$

$$\sec \alpha = \tan \alpha + \tan \left(\frac{90° - \alpha}{2}\right)$$

and so on. In the *Complete Theory of Surveying* there are also 'small tables for the eight lines cutting a circle'. These are four-figure tables for trigonometric functions, including sine, tangent, and secant, with an interval of 15 minutes. (see Table 7.1).

In Chapter 6 of the *Chóng Zhēn reign Treatise on (Astronomy and) Calendrical Science* there is a Table of Trigonometric Functions (割圓八綫表 , Gē tú bā xiàn biǎo, literally 'Table for the eight lines cutting a circle', 1631 AD): there are five-figure trigonometric tables, the interval is one minute, with values calculated by linear interpolation (see Table 7.2).

Because of the demands of astronomical calculations the methods of spherical trigonometry occupy a very important place in these translations. The *Complete Theory of Surveying* and the writings of Smogułęcki and Xuē

Table 7.1

	sin	tan	sec
0° 0′			
15′	0.0043	0.0043	1.0000
30′	0.0087	0.0087	1.0000
45′	0.0130	0.0130	1.0001
1° 0′	0.0174	0.0174	1.0001
15′	0.0218	0.0218	1.0002
30′	0.0261	0.0262	1.0003
. . .	. . .	. . .	. . .

Table 7.2

0°	sin	tan	sec	cos	cot	csc	
0'	00000	00000	1.00000	1.00000			60'
1'	.00029	.00029	1.00000	.99999	3437.74667	3437.74682	59'
2'	.00058	.00058	1.00000	.99999	1718.87319	1718.87348	59'
3'	.00087	.00087	1.00000	.99999	1145.91530	1145.91574	57'
.	.	.	.	.	.	.	.
.	.	.	.	.	.	.	.
.	.	.	.	.	.	.	.
.	.	.	.	.	.	.	.
29'	.00844	.00844	1.00003	.99996	118.54018	118.54440	31'
30'	.00873	.00873	1.00003	.99996	114.58865	114.59301	30'
	cos	cot	csc	sin	tan	sec	89°

Fèngzuò all record methods of spherical trigonometry. A lot of formulae of spherical trigonometry came into China at this time. However, because of limitations of space we cannot go into them here.

Smogulęcki and Xuē Fèngzuò's works also brought in the methods for calculating the sides and angles of triangles by using logarithms, so they also introduced logarithmic trigonometric functions (details below).

Logarithms

Logarithms were invented by the Scottish mathematician John Napier (1550–1617 AD) and were first published in a paper in 1614. This invention, analytical geometry and the calculus, are well-known as the three most important achievements of 17th century mathematics.

In 1653 AD the missionary Smogulęcki taught Xuē Fèngzuò logarithms. There is a brief introduction to logarithms in their joint work *Logarithm Tables with Explanations* (比例對數表 , *Bǐ lì duì shù biǎo*). At that time they were not called 'logarithms' but 'corresponding numbers' (比例數 , *Bǐ lì shù*) or 'power numbers' (假數 , jiǎ shù). In Chapter 12 of the book *Logarithm Tables with Explanations*, six-figure tables of logarithms for 1 to 10 000 are given, for example:

Number	Corresponding number (logarithm)
1	0.000000
2	0.301030
3	0.477121
4	0.602060
...	...

Logarithms were originally brought in for their convenience in astronomical calculations. The calculations in various books written by Smogulęcki and Xuē Fèngzuò were done using logarithms. They said: 'Change multiplications and divisions into additions and subtractions' and 'it saves six or seven tenths of the work compared with the earlier procedures and in addition to that there is no worrying about errors in multiplications and divisions'. In the book *Essentials of Trigonometry* (三角法要 , *Sān jiǎo fǎ yào*) they introduced general methods for various types of logarithmic trigonometrical calculations. For example, the sine rule

$$\frac{a}{\sin A} = \frac{b}{\sin B} = \frac{c}{\sin C} \, ,$$

was changed to

$$\log b = \log a + \log \sin B - \log \sin A$$

for calculation, etc.

Later on, in the *Collected Basic Principles of Mathematics* (1723 AD) edited by the Emperor Kāng Xī, there were more detailed discussions of logarithms. Before this book came out logarithms had no great influence on Chinese mathematics and people using logarithms were rare.

Proportional dividers, Napier's bones, and the Western calculating rules

'Proportional dividers', also known as 'Galilean proportional dividers' were introduced in a book written in 1606 AD by the great Italian scientist Galileo (1564–1642 AD). The missionary Giacomo Rho introduced proportional dividers into China in his book *Manual for Proportional Dividers* (比例規解 , *Bǐ lì guī jiě*) (a volume in the *Chóng Zhēn Reign Treatise on (Astronomy and) Calendrical Science* of 1631 AD). The proportional dividers brought into China are today in the Běijīng museum. Most of them are made of bronze or ivory, and some of them were made in China.

Proportional dividers are shaped like compasses. There are two types, with either pointed or blunt legs. Various graduations are inscribed on the two legs as in Fig. 7.7. Proportional dividers were constructed using the principle of

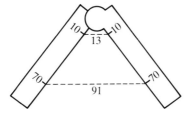

Fig. 7.6 Using proportional dividers to multiply.

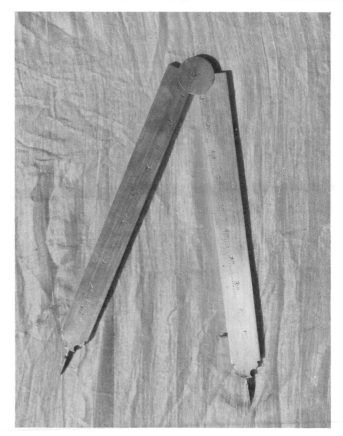

Fig. 7.7 Galilean dividers. (The original is in the Imperial Palace Museum.)

comparing the corresponding sides of similar triangles and they may be used for various sorts of calculation such as multiplication, division, finding the middle term of a ratio, extracting square roots, extracting cube roots, etc.

For example, 7×13 can be calculated using the following procedure (see Fig. 7.6). First locate the 10 graduation on one leg of the proportional dividers. Second, open the dividers so as to make the distance between the 10 graduations on the two legs be 13, that is, the base of the isosceles triangle formed by the two legs has length 13. Then locate the 70 graduations on the legs and measure the length of the base between the 70 graduations on the two legs thus getting the answer $7 \times 13 = 91$. This sort of method of calculation can be verified using the theorems on the proportions of the sides of similar triangles.

Extracting square or cube roots requires using other sorts of graduations. So on the ordinary dividers there were four or five rows of different kinds of graduations along the two legs.

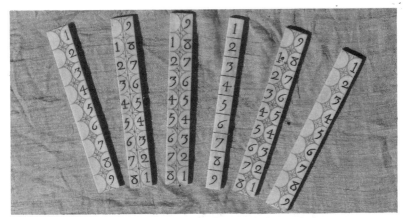

Fig. 7.8 Napier's bones. (The original is in the Imperial Palace Museum.)

After proportional dividers came in during the end of Míng and the beginning of Qīng they were very popular with scholars who were looking into Western mathematics.

Another type of calculating device brought into China during the end of Míng and the beginning of Qīng was the 'Western counting rods' (西洋籌算 , Xī yáng chōu suàn). Because the Scottish mathematician Napier (the inventor of logarithms) had written a book introducing them, this sort of calculating device was also known as 'Napier's bones'. Napier's bones were introduced by Giacomo Rho in the book *Napier's Bones* (籌算 , *Chōu suàn*, 1628 AD), which was included in the *Chóng Zhēn Reign Treatise on (Astronomy and) Calendrical Science*.

Napier's bones (see Fig. 7.8) are equivalent to a sort of separable multiplication table. Using Napier's bones multiplications and divisions can be changed into additions and subtractions. Take 85714 × 1260 as an example. The method of calculation is as follows. Take out the bones 8, 5, 7, 1, and 4 and line them up as on the right. Then take rows 1, 2 and 6: this is like the method mentioned above of 'calculating on the ground'

	8	5	7	1	4	
	8	5	7	1	4	1
1	6	0	4	2	8	2
2	4	5	1	3	2	3
3	2	0	8	4	6	4
4	0	5	5	5	0	5
4	8	0	2	6	4	6
5	6	5	9	7	8	7
6	4	0	6	8	2	8
9	2	5	3	9	6	9

(what is different is that we do not have to draw squares any longer and it is not necessary to multiply individually).

Now add up the numbers in the diagonals along each row thereby obtaining

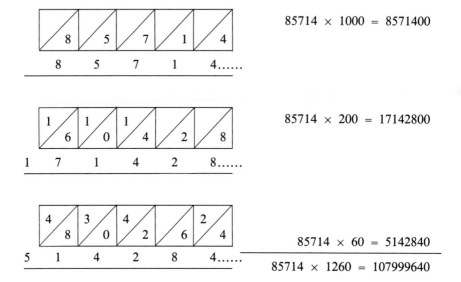

85714 × 1000 = 8571400

85714 × 200 = 17142800

85714 × 60 = 5142840

85714 × 1260 = 107999640

The Western rules and calculating machines were also introduced and may be seen in Figs. 7.9 and 7.10. These instruments can still be seen to the present day in the Imperial Palace Museum in Běijīng. Among the Western rules there were several types: the 'logarithmic rule', the 'log sine rule', and the 'log tan rule', all belonging to the early generation of calculating rules. They are the same as the so-called Gunter calculating ruler (Gunter was an Englishman, 1581–1626 AD) and there was still no cursor on the rule nor did it yet have a slide. According to research the calculating machines coming in were of the type invented by Pascal (French, 1623–1665 AD) in 1642 and were the distant ancestors of the recent (non-electric) hand calculators.

7.4 Méi Wéndǐng (梅文鼎) and his mathematics

After Western mathematics had come into China at the end of the Míng (mainly during the first 40 years of the 17th century), various mathematical works written by Méi Wéndǐng (1633–1721 AD) appeared around the beginning of the Qīng Dynasty. In the 18th century there was the *Collected Basic Principles of Mathematics* (published 1723 AD) compiled by the Emperor Kāng Xī. These works indicate that after the initial stages of the first introduction of Western mathematics, the mathematicians in China at that time

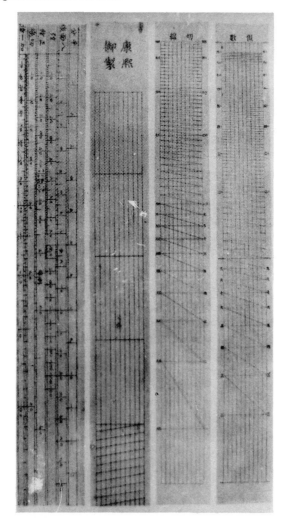

Fig. 7.9 Calculating rule. (The original is in the Imperial Palace Museum.)

were able to absorb the various sorts of mathematical methods introduced and to digest the mathematical knowledge passed into China. From this knowledge they undertook further investigations.

Méi Wéndǐng (梅文鼎), literary name Dǐngjiǔ (定九), also known as Wù'ān (勿菴) was a native of Xuānchéng (宣城) in Ānhuī (安徽), province. He was born at the end of Míng and grew up during the beginning of Qīng. This was in the initial stage of Western mathematics entering China. He did not start learning calendrical computation or mathematics until he was 27 years old, but at 33, while he was in Běijīng taking the civil service

Fig. 7.10   Calculating machine. (The original is in the Imperial Palace Museum.)

examinations he bought texts on Western calendrical computation and when
he was 42 he bought the *Chóng Zhēn reign Treatise on (Astronomy and)
Calendrical Science*. Also Méi Wénding was recommended by a friend to
participate in the work of writing the *Book of the Calendar* in the *History of
the Ming Dynasty* (明史 · 曆志 , *Ming shi, li zhi*). He spent his whole life
studying mathematics and calendrical computation; time flew past, but he
did not take up any official position. In his lifetime he wrote more than 80
works. Many of Méi Wénding's descendants knew mathematics and his
grandson Méi Jué-chéng (梅瑴成) was invited to the palace by the Emperor
Kāng Xī, who learnt mathematics from him. He also took part in the editing
of the *Collected Basic Principles of Mathematics*.

    In 1761 AD, 40 years after the death of Méi Wénding, Méi Juéchéng
compiled his written commentaries into the *Collected Works of the Méi
Family* (梅氏叢書輯要 , *Méi shì cóngshū jiyào*). Unfortunately there is no
Western language translation of this book but Martzloff (1981) has published
a detailed study of Méi Wénding's work in French. In it are collected the
works of Méi Wénding on mathematics, astronomy, and the computation of
the calendar. The parts concerning mathematics are:

    1. *Pen calculations (*筆算 , *Bisuàn*), five chapters: introducing Western
pen and paper calculations.
Appendix: The 'Fast method for measuring fields' (方田通法 , Fāng tiān
tōng fǎ) — quick methods for calculating the areas of fields and including
verses for abacus calculations. 'Investigations on ancient calculating devices'
(古算器考 , Gǔ suàn qì kǎo), an essay examining the counting rod calculations of ancient China.

2. *Napier's bones* (籌算 , *Chōu suàn*), two chapters: introducing Napier's bones.

3. *Proportional dividers* (度算釋例 , *Dù suàn shì lì*), two chapters: introducing Galilean proportional dividers.

4. *Supplement to 'What Width?'* (少廣拾遺 , *Shǎo guǎng shi yí*), one chapter: introducing the methods for extracting higher degree roots from ancient China (the highest is a 12th root).

5. *Theory of rectangular arrays* (方程論 , *Fāng chéng lùn*), six chapters: introduces the method of solution of systems of linear equations from ancient China.

6. *Right-angled Triangles* (勾股舉隅 , *Gōugǔ jǔ yú*), one chapter: contains problems on right-angled triangles, the relations between the three sides, their sums and differences.

7. *Explanations in Geometry* (幾何通解 , *Jǐhé tōng jiě*), one chapter: contains solutions using Pythagoras' [Gōugǔ] Theorem of problems concerning Books 2, 3, 4, and 6 of Euclid's *Elements of Geometry*.

8. *Elements of Plane Trigonometry* (平三角舉要 , *Píng sān jiǎo jǔ yào*), five chapters: contains plane trigonometry.

9. *Squares and Circles, Cubes and Spheres* (方圓冪積 , *Fāng yuán mì jī*), one chapter: introduces problems on inscribed and circumscribed circles and squares, and inscribed and circumscribed spheres and cubes.

10. *Supplement to Geometry* (幾何補篇 , *Jǐhé bǔ biān*), four chapters: discusses problems on regular tetrahedra, octahedra, and regular solids. In the preface it says: 'This is to supplement Euclid's *Elements*'.

11. *Elements of Spherical Trigonometry* (弧三角舉要 , *Hú sān jiǎo jǔ yào*), five chapters: contains techniques of spherical trigonometry.

12. *Geodesy* (環中黍尺 , *Huán zhōng shǔ chǐ*), five chapters: basically on the proofs of geometric theorems concerning cosines of angles in spherical triangles.

13. *Surveying Solids* (塹堵測量 , *Qiàndǔ cèliáng*): basically on geometric proofs concerning the relationship between right angled triangles on spheres and spherical angles. Qiàndǔ (塹堵) is one of the special solids discussed in the Nine Chapters (see p. 43 above; (f)).

Besides these there are some surviving manuscripts by Méi Wéndǐng that were not published.

Méi Wéndǐng's works bear on almost all aspects of the Western mathematics which had come in at that time and moreover, he was not just collecting, he also did the initial task of assimilating it and some further research. Although most of the books were based on the translation of Euclid's *Elements of Geometry* and the *Treatise on European Arithmetic* as well as the *Chóng Zhēn Reign Treatise on (Astronomy and) Calendrical Science*, he had already understood and blended these together: he was not just copying. All his work was presented in his own language.

In particular, compared with the sections in the *Chóng Zhēn Reign*

Treatise on (Astronomy and) Calendrical Science the treatment of the methods of plane and spherical trigonometry is orderly in his *Elements of Plane Trigonometry* and *Elements of Spherical Trigonometry* and in fact these two books are relatively good introductory texts. Méi Wéndĭng was well-versed in trigonometry and there is a very close relationship between his work and methods for computing the calendar.

Méi Wéndĭng not only systematically treated, edited, and described the mathematics that came in, he himself could augment it and present something extra. For example some of the problems in the *Supplement to Geometry* and *Surveying Solids* are not contained in the first six books of the *Elements of Geometry*, which had been translated into Chinese. He calculated the volumes of regular solids with twelve surfaces, twenty surfaces, etc.

Although Méi Wéndĭng's work mentions the mathematics of ancient China, because of various limitations at that time not only was he unable to get hold of the 'Ten Books of Mathematical Classics' (十部算經 , *Shí bù suànjīng*) from ancient times, he was also unable even to read the works of the Sòng and Yuán mathematicians such as Qín Jiǔsháo, Zhū Shìjié, etc.; this is a pity.

The ancient texts of China Méi Wéndĭng did see were written in columns, so he wrote the Western mathematical formulae in columns. This sort of change is unnecessary.

Méi Wéndĭng lived at the end of Míng and the beginning of Qīng, he was learned in Western mathematics and he opened the way to research into the introduced mathematics. He is the first example of someone assimilating Western mathematics and presenting it: it may be said that he is a key figure, receiving from his forerunners and opening the way for his successors. His work had a very great influence on the later compilation of the collection of works *Collected Basic Principles of Mathematics*. Ruǎn Yuán of the Qīng Dynasty described his work in the *Biographies of (Chinese) Mathematicians and Astronomers* thus: 'in his presentation of mathematical theories his aim was clarity, his sentences are neither clumsy nor wordy, he usually used simple language to explain difficult methods, everyday language to make the deepest reasoning accessible, . . . all of these were good intentions'. Méi Wéndĭng used a simple and easily understood presentation to describe what most people would regard as difficult mathematical problems, this was his most valuable and rare quality and these are the points we should learn from.

7.5 Emperor Kāng Xī (康熙) and the *Collected Basic Principles of Mathematics*

Emperor Kāng Xī (1654–1722 AD) was a Manchu originally named Àixīnjuéluó (in Chinese transcription, 愛新覺羅), who took as his Chinese name Xuán yè (玄燁). He was the second emperor of the Qīng Dynasty after the penetration of the Great Wall by the Manchus. He played a very important role in strengthening the rule of the Royal House of Qīng and restoring the national economy in the days after the devastation of the war. Emperor Kāng Xī showed intense interest in the mathematical sciences including astronomy, the computation of the calendar, and mathematics. He himself spent a considerable amount of time learning mathematics and also commanded people to translate a number of mathematical texts. It is quite a rarity in Chinese history for an emperor in a feudal system to learn mathematics and other sorts of natural science.

In Western Europe at that time France was daily growing in power. The French king Louis XIV was anxious to extend his influence in the Far East in order to counteract the growing influence of Portugal. He sent a lot of missionaries to China for various sort of activities. Amongst them those who were well-acquainted with astronomy, calendar computation and mathematics were the Jesuits:

Jean-Francois Gerbillon (1654–1707 AD, French) arrived in China in 1687 AD and Joachim Bouvet (1656–1730 AD, French) arrived in China in 1687 AD.

At that time, because Gerbillon and Bouvet were well versed in mathematics, they were asked to stay in the capital for the 'purpose of serving the Court' and they became teachers of Emperor Kāng Xī.

As far as Emperor Kāng Xī's learning mathematics is concerned, we can see the overall situation from the material presented below. In the *Correct [Catholic] Religion Receiving Praise* (正教奉褒 , *Zhēng jiào fēng bāo*) there is the following passage:

28th year of the Kāng Xī reign (1689 AD.), 25th day of the 12th moon, his Celestial Majesty [that is Emperor Kāng Xī] summoned Xú Rishēng (徐日升) [Thomas Pereyra], Gerbillon, Bouvet, An Duō (安多) [Antoine Thomas — all four were foreign missionaries: Pereyra was a Portuguese and the others French Jesuits] into the Inner Palace (Nèi Tíng, 内廷). The Emperor ordered them from then on to take turns each day in the Yǎng Xīn (养心) Hall, to lecture in Manchu [the official Court language] on Western science such as surveying etc. to His Celestial Majesty. Whenever his Majesty was free, he concentrated on learning. He loved the various disciplines: surveying, mensuration, calculating, astronomy, geometry, and logical argument. Whether he was in the capital or in the Summer Palace or touring the provinces he had to have Gerbillon and the others in his retinue. Either every day or on alternate days they lectured on Western knowledge; at the same time he ordered the lecture notes used in the Palace to be translated into the Manchu language. . . . For several years Gerbillon and the others lectured to His Celestial Majesty who rewarded them all the time.

According to reports translated by later people Gerbillon said:

Every morning we arrived in the Inner Palace at four o'clock in the morning to wait on His Celestial Majesty and even after sunset we were still not allowed to return to our residence. Every day from two hours before noon to two hours after noon we were at the Emperor's side lecturing on Euclid's Geometry or physics or astronomy, etc. We also demonstrated the methods for the computation of the calendar and the method of aiming cannon. After returning to our residence we had again to prepare for the next day's work until late in the night before retiring. It was often like this.

Even now there are several hand-written mathematical texts still preserved in the Běijīng Palace Museum that were used as lecture material. These manuscripts were written in Manchu; they were compiled from the lecture notes of the missionaries. (For example, Gerbillon's lectures, to Kāng Xī on geometry were based on the book *Elémens de géometrie où, . . .* by the Frenchman I.–G. Pardies (1671) which had been made into Manchu lecture notes. They lack the parts concerning the comparison of similar shapes, and Pythagoras' famous theorem was not listed. In addition there were the *Elements of Geometry* in Chinese, the *Outline of Essentials of Calculation* (算法纂要總綱 , *Suànfǎ zuǎn yào zǒng gōng*), *Highlights of the Method of Calculation for Extracting Roots* (借根方算法節要 , *Jiè gēn fāng suànfǎ shāi yào*), the *Theory of Solving Triangles* (三角形推算法論 , *Sān jiǎo xing tuī suàn fǎ lùn*), the *Manual on the Theodolite* (測量高遠儀器用法 , *Cè liáng gāo yuǎn yi qi yòng fǎ*) the *Explanation of the Proportional Dividers* (比例規解 , *Bǐ lǐ guī jiě*) and *Trigonometrical Tables* (八綫表根 , *Bā xiàn biǎo gēn*). These works formed the reference material for the compilation of the *Collected Basic Principles of Mathematics*.

In addition to the foreign missionaries, Emperor Kāng Xī summoned the men in the country who at that time were well-versed in mathematics, such as Méi Wéndǐng and Chén Hòuyào (陳厚耀), to discuss mathematical problems and later he also summoned Méi Wéndǐng's grandson Méi Juéchéng to teach mathematics in the Court. Emperor Kāng Xī himself taught geometry to the royal princes inside the palace.

The compilation of the *Collected Basic Principles of Mathematics* was another important event supervised by Emperor Kāng Xī. In the 51st year of the Kāng Xī reign (1712 AD) Chén Hòuyào put a proposal to the Emperor Kāng Xī 'to produce a definitive edition of the mathematical texts in order to benefit the country'. Emperor Kāng Xī immediately put Méi Juéchéng in charge with Chén Hóuyào, Hé Guózóng (何國宗), Míng Āntú (明安圖), and others to do the compiling. After almost 10 years of work, 100 chapters of the *Ocean of Calendrical and Acoustical Calculations* (律曆淵源 , *Lü lì yuān yuán*) were ready in the 60th year of the Kāng Xī reign (1721 AD). It is divided into three parts:

Compendium of Calendrical Science and Astronomy (曆象考成 , *Lì xiàng kǎo chéng*), 42 chapters (on astronomy and the rules for calendars);

Collected Basic Principles of Mathematics (數理精蘊 , *Shù lǐ jīng yùn*), 53 chapters (on various sorts of mathematics); and

Collected Basic Principles of Music (律呂正義 , *Lü lü zhēng yì*), 5 chapters (on music theory).

This book was printed in the first year of the Yōng Zhēng (雍正) reign (1723 AD), but by that time the Emperor Kāng Xī was dead.

The *Collected Basic Principles of Mathematics* took the Western mathematical knowledge that had been newly introduced into China and treated it in an orderly and logical sequence: it started the comparative study of mathematics in ancient China. However this was confined to the mathematical knowledge available in the texts circulating at that time as some important books were then lost. The books covered all aspects of mathematical knowledge at that moment and therefore can be regarded as a mathematical encyclopaedia representing the level of mathematics then. Because it bore a title decreed by the Emperor Kāng Xī it was distributed throughout the country, so its circulation was very wide and its influence very great. Because of the circumstances its circulation was much wider than that of the *Elements of Geometry* (translated by Xú Guāngqī) and the *Treatise on European Arithmetic* by Matteo Ricci (translated by Lǐ Zhīzǎo). This book remained a compulsory text for learning mathematics for a very long time but it was also a very important reference book for mathematical research.

The *Collected Basic Principles of Mathematics* is divided into two volumes. The contents of the first volume in five chapters are 'to establish the objectives and to understand the system'; the 40 chapters of the second volume are divided into 'specific parts and for applications'; and in addition there are four types of tables contained in eight chapters.

The first chapter of the first volume of the *Collected Basic Principles of Mathematics* is the 'Source of mathematics' and in explaining the source of numbers it still copies the 'hé-tú and luò-shū' diagrams and its approach is based on number mysticism. Following that it also uses a passage from the *Zhōubǐ Suànjīng*. The second, third, and fourth chapters were called the 'Elements of geometry' (幾何原本 , Xiàn hé yuán běn) and were based on the revised lecture notes of the missionary Gerbillon. The proofs of the theorems and the order are significantly different from those of Euclid's *Elements of Geometry* and very different from Xú Guāngqī's translation. In fact the *Elements of Geometry* in the *Collected Basic Principles of Mathematics* is a translation of a particular geometry textbook that was very popular in France in those days.

Chapter 5 of the first volume, entitled the 'Source of computation methods' (算法原本 , Suàn fǎ yuán běn) discusses the multiplication of natural numbers, common multiples, common divisors, ratios, and arithmetic and geometric series but it includes neither prime nor co-prime numbers among its contents.

The second volume of the *Collected Basic Principles of Mathematics* is divided into five large sections: 'Introduction', 'Lines', 'Surfaces', 'Solids', and 'Conclusion'. Here is a brief survey of these various parts:

Introduction (two chapters): describes the systems of measurement of lengths, weight, and content, the systems for fixing the decimal place, and the four arithmetic operations for integers and fractions.

Lines (eight chapters): contains problems on various sorts of ratio, the method of calculation for excess and deficit, and the technique of 'rectangular arrays' — that is the method of solution of systems of linear equations from ancient China.

Surfaces (ten chapters): various sorts of problems on triangles (including problems on surveying), areas of various rectilinear figures, areas of circles and their segments, ellipses, problems on extracting roots and the corollary to extracting roots, etc.

Solids (eight chapters): various volumes (including spheres, segments of spheres, ellipsoids, etc.), the calculation of the volumes of various regular solids, and also the lengths of the sides of various sorts of regular solid and their relations with their diameters of circumscribed and inscribed spheres. Problems on extracting cube roots and the corollary to extracting cube roots, etc.

Conclusion (ten chapters): contains 'the completing the square solution of quadratic equations' (借根方比例 · Jiè gēn fāng bǐ lì). This is the Western algebra that had been introduced into China by then. In addition it also treats logarithms and Galilean proportional dividers.

Most of the various sorts of calculation in the *Collected Basic Principles of Mathematics* do not go beyond the limits of the various mathematics books compiled by Méi Wéndǐng or the *Treatise on European Arithmetic*, but towards the end of the 'Conclusion' there are some new items that had not previously appeared. These are the method of computing tables of logarithms and the introduction of the completing the square method of solution of quadratic equations.

As mentioned above, the use of logarithms had already been introduced by the work of the missionary Smogulęcki and Xuē Fèngzuò. However the method for computing these tables of logarithms was recorded for the first time in the *Collected Basic Principles of Mathematics*.

In chapter 38 of the *Collected Basic Principles of Mathematics*, *Logarithms* it says: '1 is the beginning of number, when used to multiply or divide, the number does not change, so the logarithm of 1 is defined to be 0. The logarithm of 10 is 1, the logarithm of 100 is 2, the logarithm of 1000 is 3, the logarithm of 10 000 is 4, . . . similarly going on to myriads of myriads just add on 1 each time. This is the key to logarithms.' This is to say, first of all set

$$\log 1 = 0,$$
$$\log 10 = 1,$$
$$\log 100 = 2,$$
$$\log 1000 = 3,$$
$$\cdots$$

But for the intermediate values between 1 and 10, 10 and 100 etc., how are the logarithms to be computed? Three types of method were presented in the

Collected Basic Principles of Mathematics. This is the first time the computation of logarithms was presented in China. Although in the Europe of that time Gregory and N. Mercator (1668 AD) had already used the method of series to compute such tables, nevertheless this advanced method of getting the tables did not pass into China. Of the three methods in the *Collected Basic Principles of Mathematics*, the two that were more practicable both required taking successive square roots and were very complicated. The method of series was only eventually adopted by Dài Xǔ (戴煦) and Lǐ Shànlán (李善蘭). However their methods were not influenced by the West, they were derived through their own independent thought and research (details in the next chapter).

The final part of the *Collected Basic Principles of Mathematics* contained something new, namely 'completing the square', which can be found in Chapter 31. This was the Western algebra of those times and its origin was in the Arab countries of the Middle Ages. In this the unknown number is called the 'root'. According to the explanation in the *Collected Basic Principles of Mathematics*: 'Taking the unknown square and a number of unknowns assumed equal to a given number, find the unknown number.' That is to say 'Taking the unknown square' is to take multiples of the unknown number and its square and form an equation and then solve the equation to find the unknown number. Obviously this is very similar to the technique of the celestial element in the Sòng and Yuán period in China. So Méi Juéchéng and the others claimed that the 'technique of the celestial element' was like 'completing the square'.

The technical terms and symbolism in the discussion of 'completing the square' in the *Collected Basic Principles of Mathematics* may be illustrated by giving the following example. The question says:

'Suppose there are four cubes, add on three squares, take away two roots, add on five numbers and also five cubes, take away one square, add on three roots, take away two units: add. What does one get?'

This is the addition of the polynomials

$$(4x^3 + 3x^2 - 2x + 5) + (5x^3 - x^2 + 3x = 2).$$

The method of recording this in the *Collected Basic Principles of Mathematics* is as follows:

四立方 ⌐ 三平方 —— 二根 ⌐ 五真數	4 cubes + 3 squares	– 2 roots	5 numbers
五立方 —— 一平方 ⌐ 三根 —— 二真數	5 cubes – 1 square	+ 3 roots	– 2 numbers
九立方 ⌐ 二平方 —— 一根 ⌐ 三真數	9 cubes + 2 squares	+ 1 root	+ 3 numbers

Another problem on equations says:

'Suppose a cube less nine roots is equal to 1620 feet. What is the value of the root?'

This is a cubic equation, $x^3 - 9x = 1620$; its root is 12.

In the *Collected Basic Principles of Mathematics* it says:

一立方——九根＝一六二〇 ，   1 cube – 9 roots = 1620,
一根＝一二 。   1 root = 12.

Here Chinese figures for 1, 2, . . ., 9, 0 are being used in the Arabic way to represent numbers bigger than ten.

In the 'completing the square' recorded in the *Collected Basic Principles of Mathematics*, rational functions are introduced: the addition, subtraction, multiplication, and division of polynomials, the method for extracting roots, and the corollary to extracting roots (quadratic equations and general equations including the method of solution for higher degree equations) and other sorts of practical problems.

Completing the square was also then called 'Algebra' (阿爾熱巴拉 , Ā ěr rè bā lā) — this is a phonetic transcription.

A little later Descartes' notation was also introduced into China and recorded in *New Methods in Algebra* (阿爾熱巴拉新法 , Ā ěr rè bā lā xin fǎ — two chapters, manuscript: Lǐ Yǎn had a reproduction of it). In it the Ten Heavenly stems (甲 , 乙 , 丙 , 丁 , etc.) were used to represent given numbers a, b, c, . . . and the twelve Earthly Branches (子 , 丑 , 寅 , 卯 , etc.) to represent unknown numbers x, y, z, . . . (see above, p. 22). □ was used to denote plus and □□ to denote minus (they are symbols used in Yīn–Yáng fortune telling) and   was used to denote equality. For example

亥亥 □ 甲丙卐乙亥 ，
亥亥亥□□甲亥卐庚 ，

represents,

$$zz + ac = bz,$$
$$zzz - az = k .$$

8

Mathematics under the feudalistic closed door policy in the middle period of the Qīng

8.1 The change in the academic environment

The first importation of Western mathematics, from the end of Míng in the last years of the 16th century until the compiling and printing of the *Collected Basic Principles of Mathematics* (1723 AD) at the end of the Kāng Xī Reign, lasted altogether for about a century and a half. The mathematical knowledge that entered during this period was sketched in the previous chapter. In the Yōng Zhēng (雍正 , 1723–1735 AD) reign the Qīng government adopted a closed-door policy; work on the propagation and translation of Western mathematics stopped and at that point the history of mathematics in the Qīng period again entered a new era.

As early as the end of Míng and the beginning of Qīng when the Western mathematics entering was gradually increasing, while the two governments of the Míng and Qīng and the educated classes of the time were absorbing the achievements of Western science, they simultaneously were fearful of the missionaries winning over the hearts of the populace and thereby threatening the working of the feudalistic regime, so they were on their guard against the evangelization and other activities of the missionaries. In the year 1704 (the 43rd year of the Kāng Xī reign) the Pope in Rome decreed that Chinese catholics were forbidden to pay respects to their ancestors according to their traditions and customs. This was the result of the famous Rites Controversy. (See Latourette 1929/1966, pp. 140ff.) This order caused an outrage both in the Court and in the Chinese populace. Consequently in the 46th year of his reign, Emperor Kāng Xī ordered the papal legate, de Tournon, to be transferred and confined in Macao (see Rowbotham 1966, Chapter XI.)

Then the missionaries made plans, by various means, to help the princes who were sympathetic to the Catholics succeed to the throne. Therefore, when another prince, who was supported by the Buddhist lamas, succeeded to the throne and became the Yōng Zhēng emperor, all the foreign missionaries were immediately deported to Macao; They were not allowed to travel freely into the mainland and only a few remained to work in the State Observatory. At the same time the Chinese people were forbidden to leave the country and overseas Chinese were forbidden to return. This sort of feudalistic closed-door policy continued until after the Opium Wars (1840 AD) in the Dào Guāng (道光) reign, lasting altogether for more than 120 years. During this period almost no new Western mathematical knowledge entered China. Consequently Chinese mathematicians concentrated their efforts on the mathematics that had previously come in, carefully digested it, and proceeded to conduct further research.

In this period of the history of mathematics there was another important achievement, namely a large-scale study and collating of the mathematics of ancient China. This was bound up with the popularity of the so-called 'Qián Jiā school' (乾嘉學派, Qián Jiā xuépài) of that time, that is, the 'textual criticism' (漢學家, Hàn xué jiā) of the 'Sinologists' (考據學, Kǎo jù xué). At that time a lot of scholars also used the methods of textual criticism to revise the ancient mathematical texts.

One way the Qīng government treated the educated was to put a lot of pressure on them and to jail many for their writings. Occasionally one or two words or one or two lines of a poem could cause the calamity of all members of the nine degrees of kindred (i.e. almost any relative) being put to death. On the other hand, the government established the 'Institute for compiling the *Complete Library of the Four Branches of Literature*' (四庫全書館 , Sì kù quān shū guǎn), which engendered a large-scale collating of traditional Chinese texts. Those books disliked by the rulers were either forbidden or immediately burnt and destroyed. On the one hand they imposed severe restrictions and on the other they inveigled people into working for the government. The 'textual criticism school' and the 'Sinologists' flourished under this encouragement. It became the prevailing attitude to be pre-occupied with examining, commenting on, and researching into the various copies of the ancient works that had survived. There were quite a number of people who started work on collecting, collating, examining, commenting on, and reprinting books on certain aspects of science, including books, and manuscripts on mathematics.

So research into and editing of the ancient Chinese mathematical texts and the digesting and further investigation of the Western mathematical knowledge that had previously come in became the two major areas in the mathematics of this period. A lot of mathematicians achieved very good results in these two areas and the most well-known among them are the following:

Chén Shirén (陳世仁, 1676–1722),
Míng Āntú (明安圖, ? –1765),
Lǐ Huáng (李潢, ? –1811),
Jiāo Xún (焦循, 1763–1820),
Wāng Lái (汪萊 , 1768–1813),
Lǐ Rui (李銳, 1773–1817),
Xiàng Míngdá (項名達1789–1850),
Shěn Qīnpéi (沈欽裴, ? – ?),
Luó Shilín (羅士琳 , 1789–1853),
Dǒng Yòuchéng (董祐誠 , 1791–1823),
Dài Xǔ (戴煦, 1805–1860),
Lǐ Shànlán (李善蘭 , 1811–1882).

The achievements of these people in the work of collating the ancient Chinese mathematics are described briefly below.

8.2 The collating of ancient Chinese mathematical texts

The work on the collating of the ancient Chinese mathematical texts can be divided into four parts, namely:

1. The editing of various collections of books.
2. The collating of the *Ten Mathematical Manuals* (算經十書 , *Suànjīng shí shū*).
3. The collating of the Sòng and Yuán mathematical texts.
4. The compilation of the *Biographies of Mathematicians and Astronomers* (疇人傳, *Chóu rén zhuàn*).

1 The editing of various collections of books

As well as the *Collected Basic Principles of Mathematics* (as part of the *Ocean of Calendrial and Acoustical Calculations*) at the end of the Kāng Xī reign, another even bigger collection of books in the style of an encyclopaedia was compiled, entitled *Collection of Ancient and Modern Books* (古今圖書集成 , *Gǔ jīn tú shū jí chéng*) with 10 000 chapters. This collection of texts was completed in the fourth year of the Yōng Zhēng reign (1726 AD), the moveable type was set up and printed in the Hall of Military Heroes (武英殿 , *Wǔ yīng diàn*) and the collection was called *Treasure Accumulating Editions* (聚珍版 *Jù zhēn bǎn*) printed in movable type. Some of the Western calendrical and mathematical texts in the *Treatise on Mathematics (Astronomy and Calendrical Science) According to the New Methods* (新法曆書, *Xīn fǎ lì shū*) were included; as to ancient Chinese mathematical works, only five books were included, *Zhōubì suànjīng, Memoir on some Traditions of Mathematical Art* (數術記遺 , *Shù shù jì yí*), the *Mathematical Manual of Xiè Cháwēi* (謝察微算經 , *Xiè Chá wēi suànjīng*), the mathematical part of *Dream Pool Essays* (夢溪筆談, *Mèng qí bǐ tán*) and the *Systematic Treatise on Arithmetic* (算法統宗, *Suàn fǎ tǒng zōng*).

The largest collection of books on the Qing Dynasty has to be the *Complete Library of the Four Branches of Literature*. The first set of the *Complete Library* was completed between the 38th year of the Qián Lóng (乾隆) reign (1773 AD) and the winter of the 46th year of the Qián Lóng reign (1781 AD). This first set of the *Complete Library* has altogether 3459 divisions with 36 078 volumes. There were seven sets of the *Complete Library* but the number of volumes varied from set to set. Besides Dài Zhèn (戴震) among those who were well-versed in mathematics and therefore participated in the work of compilation were Chén Jìxin (陳際新), Guō Zhǎngfā (郭長發), and Ní Tíngméi (倪廷梅).

Books of the 'Category of astronomy and calculation methods' were collected in Part A, Sections 16 and 17 of the *Complete Library* and according to the record in the *Summary of the Index of the Complete Library* (四庫全書總目提要 , *Sì kù quán shū zǒng mu tí yào*), this class had 58 divisions with 579 volumes. In the course of the compilation of the *Complete Library*, books were collected together throughout every principality and county from book collectors and old lost texts were copied from the *Great Encyclopaedia of the Yǒng Lè Reign* (永樂大典 , *Yǒng lè dà diǎn*). Thus a lot of rare and unknown ancient mathematical texts reappeared. The famous *Ten Books of Mathematical Classics* and the works of Qín Jiǔsháo, Lǐ Zhì, etc. of the Sòng and Yuán periods reappeared because of this collating.

The work of compiling the *Complete Library* led the mathematicians of that time into furthering their research and the collating of ancient Chinese mathematical texts.

At the same time, during the compilation of the *Complete Library*, there was also the printing of the *Imperial Collection of Treasure Accumulating Editions* (武英殿聚珍版叢書 , *Wǔ yīng diàn jù zhēn bán cóng shū*). This collection of books, printed in moveable type, contained selections of texts taken from those collected for the *Complete Library*. Later various places such as Jiāngsú (江蘇), Zhèjiāng (浙江), and Jiāngxī (江西) imitated this, reprinting the *Collection of Treasure Accumulating Editions*. In this collection of books they selected seven books from the *Ten books of Mathematical Classics* such as *Zhōubǐ*, the *Nine Chapters on the Mathematical Art* etc. and also 100 chapters from the *Ocean of Calendrical and Acoustical Calculations*.

As well as the above-mentioned government supervised collection of books, there were also printed independently the *Collected Works of the Ripple Pavilion* (微波榭叢書 , *Wēi bō xiè cóng shū*), the *Collected Works from the Zhī bù Zu Private Library* (知不足齋叢書 , *Zhī bù zú zhāi cóng shū*) and the *Collected Works of Yí Jià Hall* (宜稼堂叢書 , *Yí jià táng cóng shū*). Among these there were a lot of mathematical works assembled.

We shall briefly describe these collected works in the next two sections.

2   The collating and commentary on the *Ten Mathematical Manuals* (算經十書 , Suànjīng shí shū)

The *Zhōubǐ* and the *Nine Chapters* in the *Ten Books of Mathematical Classics* are the products of the development of mathematics during the Hàn to Táng periods in China, which extended over more than a thousand years. As mentioned earlier these 10 books were fixed as the textbooks during the Táng Dynasty, and the Northern Sòng government also printed and published them (in 1048 AD; by that time Zǔ Chōngzhī's 'method of interpolation' was lost). In the time of the Southern Sòng there were new, reset printings (1213 AD or a little later). However, by the time of the Míng Dynasty, with the exception of the *Zhōubǐ*, which was still being reprinted and circulated, the other books of the *Ten Books of Mathematical Classics* were almost completely lost. It was not until the compilation of the *Complete Library* during the Qián Lóng reign of the Qīng, that most people again had the chance to see these various ancient mathematical texts.

During the time these various mathematics books were being collated they were only available from two sources. One source was directly from the Southern Sòng edition and the other was copying from the *Great Encyclopaedia of the Yǒng Lè Reign period* of the Ming.

Actually, at that time the *Ten Books of Mathematical Classics* in the Southern Sòng edition were not completely lost. During the end of Ming and the beginning of Qīng the book collector father and son Máo Jìn (毛晋) and Máo Yì (毛扆) of the Jí Gǔ (汲古) Library acquired Southern Sòng editions of seven works: *Master Sūn, Mathematical Manual of the Five Government Departments*, *Zhāng Qiūjiàn*, *Zhōubì*, *Continuation of Ancient Mathematics* and *Xiàhóu Yáng*, but the *Nine Chapters* was incomplete. During the Kāng Xī reign a manuscript copy was sent to the Qīng Court and stored in the Royal Library (Tiān Lù Lín Láng Pavilion — 天禄琳瑯閣 , Tiān lù lín láng gě), where it was not easy for ordinary people to see it. (A photographic copy of the first section of the first edition was printed by the Palace Museum in 1931.) Later on the original Southern Sòng copies of the Máo family were in turn dispersed into other collectors' hands. During this period the *Xiàhóu Yáng* and *Continuation of Ancient Mathematics* were lost, and only five works remain. At present these titles have been collected one by one into the State Libraries and have become the property of the Chinese nation.

When Dài Zhèn participated in the compiling of the *Complete Library* he copied seven works from the *Great Encyclopaedia of the Yǒng Lè Reign Period* — the *Nine Chapters*, *Sea Island*, etc., — and, from the manuscript copies of the Máo family he added in two other titles. At the same time he published them in the *Treasure Accumulating Editions*. In the 38th year of the Qián Lóng reign (1773 AD) Kǒng Jìhán (孔繼涵 , 1739–1783 AD) selected all copies of mathematical texts from the Sòng manuscripts of the Máo family and from the *Great Encyclopaedia of the Yǒng Lé Reign Period*, printed them in the *Collected Works of the Ripple Pavilion*, and called them the *Ten Mathematical Manuals* (算經十書 , Suànjīng shí shū). The various versions of the *Ten Mathematical Manuals* were reset or reprinted from this edition (e.g. the Universal Library edition (萬有文庫 , Wàn yǒu wén kù), the lithographic edition of the Hóng Bǎo zhāi, (鴻寶齋 , etc.)

The situation concerning the circulation, copying and reprinting of the old works in the *Ten Mathematical Manuals* at the beginning of Qīng is shown in the Table 8.1.

The reappearance of the *Ten Mathematical Manuals* immediately stimulated the interest of the mathematicians of that period and they all commented on and investigated these books. Among these, those who met with outstanding success were Dài Zhèn (戴震), Li Huàng (李潢), Shěn

Table 8.1

Title of book	Edition			
	Southern Sòng (Printed 1213 AD or a little later)	Máo family Jí Gǔ private library Ms. of the Song edition (1684 AD)	Complete Library of the Four Branches of Literature 1773 AD (四庫全書, Sì kù quán shū)	Kǒng edition of the Ten Mathematical Manuals (1773 AD)
Nine Chapters on the Mathematical Art (九章算術 Jiǔzhāng suànshù)	Now in the Shànghǎi library (only five chapters survive)	Now in Palace Museum	Copied from the Great Encyclopaedia of the Yǒng-lè Reign Period (永樂大典 Yǒng-lè dà diǎn)	Edition by Dài Zhèn
Zhōubǐ suànjīng (周髀算經)	Now in Shànghǎi library	Now in Palace Museum	Copied from the Great Encyclopaedia of the Yǒng-lè Reign Period	Edition by Dài Zhèn
Master Sūn's Mathematical Manual (孫子算經, Sūnzǐ suànjīng)	Now in Shànghǎi library	Now in Palace Museum	Copied from the Great Encyclopaedia of the Yǒng-lè Reign Period	Edition by Dài Zhèn
Zhāng Qiūjiàn's Mathematical Manual (張丘建算經, Zhāng Qiūjiàn suànjīng)	Now in Shanghai library	Now in Palace Museum	Copied from the Máo family Ms. copied from the Sòng edition	Edition by Dài Zhèn
Mathematical Manual of the Five Government Departments suànjīng, Wǔ cāo) 五曹算經	Now in Běijīng University Library	Now in Palace Museum	Copied from the Great Encyclopaedia of the Yǒng-lè Reign Period	Edition by Dài Zhèn

Xià Hóuyáng's Mathematical Manual (夏侯陽算經, *Xià Hóuyáng suànjīng*)	Lost	Now in Palace Museum	Copied from the Great Encyclopaedia of the Yǒng-lè Reign Period	Edition by Dài Zhèn
Continuation of Ancient Mathematics (緝古算經, *Xù gǔ suànjīng*)	Lost	Now in Palace Museum	Copied from the Máo family Ms. copied from the Sòng edition	Edition by Dài Zhèn
Arithmetic in the Five Classics (五經算術, *Wǔjīng suànshù*)	Lost long ago	Not copied	Copied from the Great Encyclopaedia of the Yǒng-lè Reign Period	Edition by Dài Zhèn
Sea Island Mathematical Manual (海島算經, *Hǎidǎo suànjīng*)	Lost long ago	Not copied	Copied from the Great Encyclopaedia of the Yǒng-lè Reign Period	Edition by Dài Zhèn
Memoir on Some Traditions of Mathematical Art (數術記遺, *Shùshù jìyí*)	Now in Beijing University Library	Not copied	Based on the book presented by the Governor of the two Jiāngs (兩江)	?

Qīnpéi (沈欽裴), and Gù Guānguāng (顧觀光). Apart from Dài Zhèn, who was compiling the *Complete Library* and so had experience in collating, comparing, and investigating the mathematical books, other people also wrote works. The most important among them are:

Lǐ Huàng (?-1811): *Careful explanation of the 'Nine Chapters on the Mathematical Art'*, *with diagrams* (九章算術細草圖説 , *Jiuzhāng suànshù xi cǎo tú shuō*)

Lǐ Huàng: *Careful explanation of the 'Sea Island Mathematical Manuals'*, with diagrams (海島算經細草圖説 , *Haǐdǎo suànjīng xi cǎo tú shuō*)

Lǐ Huàng: *Commentary on the 'Continuation of Ancient Mathematical Methods' for Elucidating the Strange [Properties of Numbers]* (緝古算經考注 , *Xù gǔ suànjīng kǎo zhù*)

Gù Guānguāng (1799-1862): *A Textual Criticism of the Zhōubǐ Suànjīng* (周髀算經校勘記 , *Zhōubǐ suànjīng xiao kān ji*).

These works are extremely helpful even now for understanding the *Ten Mathematical Manuals* of ancient China.

3 The collating and research work on the Sòng and Yuán mathematical texts

In addition to the collation, examination of, and commentaries on the *Ten Mathematical Manuals* by a number of people, a lot of the Sòng and Yuán mathematical texts were also tidied up and reprinted so as to make them available. Among them the more important books may briefly be described as follows:

1. The *Mathematical Treatise in Nine Sections* by Qín Jiǔsháo. When the Institute of the Complete Library was established during the Qián Lóng reign this book was copied out from the *Great Encyclopaedia of the Yǒng Lè Reign Period* by Dài Zhèn, compiled into the *Complete Library* and given the title of *Mathematical Treatise in Nine Sections* (數書九章 , *Shùshū jiuzhāng*). Both Jiāo Xún (焦循) and Lǐ Ruì (李鋭) did research on this book. Later Shěn Qīnpéi (沈欽裴) discovered a Míng manuscript [this is the so-called 'Zhào Qǐměi' (趙琦美) manuscript] and he also compared these versions. Shěn Qīnpei's student Sòng Jǐngchāng (宋景昌) studied these two versions; he carefully analysed and augmented them and wrote notes and comments. Then Yù Sōngnián (郁松年) published the final version in the *Collected Books of the Yí Jià Hall* (宜稼堂叢書 , *Yí jià táng cóng shū*) (1842 AD) and gave it the name *Mathematical Treatise in Nine Sections*; this is the edition usually seen now [e.g. in the editions of the *Basic Collection of Books on Chinese Learning* (國學基本叢書 , *Guó xué ji běn cóng shū*) and the *Collection of Books* (叢書集成 , *Cóng shū ji chéng*)].

2. The various mathematical books written by Yáng Huī. Yáng Huī's books were all assembled into the *Great Encyclopaedia of the Yǒng Lè Reign*

Period, but these books were not copied during the compilation of the *Complete Library*. Only part of the *Continuation of Ancient Mathematical Methods for Elucidating the Strange [Properties of Numbers]* was edited in the *Collected Books of the Zhī Bù Zú Private Library* (知不足齋叢書 , *Zhī bù zú zhāi cóng shū*) printed by Bǎo Tíngbó (鮑廷博). In 1840 Yù Sōngniān (郁松年) printed the *Collected Works of the Yí Jià Hall* (宜稼堂叢書 , *Yí jià tāng cóng shū*) and in it he included the *Detailed Analysis of the Mathematical Methods of the 'Nine Chapters'* (with a supplement) by Yáng Huī and *Yáng Huī's Methods of Computation* (six chapters), but several of the chapters were incomplete. It was discovered many years later, at the beginning of the twentieth century, that the Koreans had, in 1433, reprinted *Yáng Huī's Methods of Computation*, from the edition published in 1378 during the Hóng Wǔ (洪武) reign of the Míng Dynasty. The Korean reprints also passed over to Japan and so were preserved up to the present day (cf. Lam Lay Yong 1977, p. xvi).

3. The two books of Lǐ Zhì, *Sea Mirror of Circle Measurements* and *New Steps in Computation*. These two books were collected into the *Complete Library*. The former is based on the book in the private library of Lǐ Huàng (李潢) and the latter was copied from the *Great Encyclopaedia of the Yǒng Lè Reign Period*. After comparison, editing, and commentary by Lǐ Ruì these two books were compiled into the *Collected Works of the Zhī Bù Zú Private Library*. Of the books in circulation today, most of them were based on the reset and reprinted editions of the *Zhī Bù Zú Private Library*.

4. The *Introduction to Mathematical Studies* and the *Precious Mirror of the Four Elements* by Zhū Shìjié. These two books were not compiled into the *Complete Library*. During the Jiā Qìng (嘉慶) reign (1796–1820 AD) Ruǎn Yuán (阮元) obtained the *Precious Mirror of the Four Elements*: it was sent to Beijing to be presented to the Court as an 'uncollected book of the *Complete Library*'. At the same time he gave a copy of the manuscript to be collated by Lǐ Ruì. Later Lǐ Ruì died and the collating work was not completed. At that time there were many other people such as Xú Yǒurén (徐有壬), Shěn Qīnpéi and others working on the *Precious Mirror of the Four Elements*. It was not until the second year of the Dào Guāng (道光) reign (1822 AD) that Luó Shìlín obtained a manuscript. He finally compared and investigated this manuscript and other editions and after more than 10 years of work he wrote the *Detailed Analysis of the 'Precious Mirror of the Four Elements'* (四元玉鑑細草 *Sì yuán yù jiàn xì cǎo*). In the 14th year of the Dào Guāng reign (1843 AD) this book was printed in Yángzhōu (揚州). This *Detailed Analysis* is the edition of the *Precious Mirror of the Four Elements* usually seen.

Shěn Qīnpéi also wrote a *Detailed Analysis of the 'Precious Mirror of the Four Elements'* but it was never published. (The Běijīng library has a copy of the manuscript.) Shěn Qīnpéi's *Detailed Analysis* has a number of distinctive features and is worthy of note.

The discovery of the *Introduction to Mathematical Studies* came a little later. In the 19th year of the Dào Guāng (道光) reign (1839 AD), Luó Shìlín reprinted this book based on the newly discovered Korean reprinted edition. This Korean reprint was published in 1660 and discovered in Běijīng by Luó Shìlín (see Lam Lay Yong 1979, p. 4). Once this book had reappeared Chinese people had a relatively full knowledge of the Sòng and Yuán mathematical works.

At that time, however, the search for and collecting of old mathematical works throughout the country was not completely thorough and it was not until recently that some of the deficiencies were rectified.

4   The Compilation of the *Biographies of Mathematicians and Astronomers* (疇人傳 , *Chóurén zhuán*)

At the end of the 18th century as well as the collating and study of ancient Chinese mathematical texts there was also the compilation of the *Biographies of (Chinese) Mathematicians and Astronomers*. Ruǎn Yuán (阮元) was the chief editor of the *Biographies of (Chinese) Mathematicians and Astronomers*, but much of the actual writing was done by Lǐ Ruì (李鋭) and Zhōu Zhìpíng (周治平). In the *Guide to the Use of the Biographies of Mathematicians and Astronomers* (疇人傳凡例 , *Chóu rénzhuàn fán lì*) by Ruǎn Yuán it says: 'The edition was started in the Yǐ Mǎo (乙卯) year of the Qián Lóng (乾隆) reign and was completed in the Jǐ Wèi (己未) year of the Jiā Qìng (嘉慶) reign', so according to this the compilation of the *Biographies of Mathematicians and Astronomers* took place in the years 1795–1799 AD.

The complete work *Biographies of Mathematicians and Astronomers* has a total of 46 chapters; it records the mathematicians and calendar makers down the ages: 243 people in all with their biographies. It also has an appendix including 37 Europeans, a grand total of 280 people. The two characters 疇人 (Chóu Rén) of the title come from the *Chronicles* written by Sīmǎ Qiān and denote the hereditary officials who had particular responsibility for astronomy and calendar making (cf. Chapter 1, p. 22). In his *Guide to the Use* Ruǎn Yuán says:

On the path of learning, the most important thing is specialized knowledge. The path to mathematics is deep and wide, only specialists can distinguish true from false. The master historian [Sīmǎ Qiān, see Chavannes 1895 Vol. III, p. 326] said: 'The descendants of the Chóu Rén [the hereditary officials] were scattered', and according to Rú Chùn (如淳): 'The family occupation, passed from generation to generation, is called Chóu (疇). At the age of 23 he is given the hereditary office, each learns from his father, these are the descendants of the experts'. This is why this work was given the name *Biographies of the Chóu Rén [Mathematicians and Astronomers]* its meaning comes from here.

People and events concerned with fortune telling and number mysticism were not included in the *Biographies of Mathematicians and Astronomers*. At the same time it expressed the opinion that in calendar-making observation should be emphasized and opposed connecting calendar-making with musical theory and the *Yì Jīng* (the *Book of Changes*). This point of view is correct. At the end of some biographies in the *Biographies of Mathematicians and Astronomers* a brief commentary was appended. Although these commentaries are not always unbiased, they nevertheless have their own valuable points.

But there were also some incorrect views in the *Biographies of Mathematicians and Astronomers*. For example, the view is expressed that 'the techniques of later generations are more precise than their predecessors, this is because of collecting together the achievements of the ancients, not because the knowledge of later generations can exceed the ancients.' It also contains the opinion that 'the Western methods were in fact copied from China'. Obviously it is unreasonable to claim that without exception Western mathematics owes its origins to China.

Almost 100 years after Ruǎn Yuán, Lǐ Ruì *et al.* had finished compiling the *Biographies of Mathematicians and Astronomers* (1799 AD), at the end of the 19th century, Luó Shìlín *et al.* made amendments on three occasions. In the twentieth year of the Dào Guāng reign (1840 AD) they wrote an *Appendix* (on 12 people referring to five other entries) and the *Continuation of the Appendix* (on 20 people with reference to seven people), a total of six chapters, which with the 46 chapters compiled by Ruǎn Yuán *et al.* made a grant total of 52 chapters. Again at the end of the Guāng Xù (光緒) reign (1886 AD) Zhū Kěbǎo (諸可寶) wrote the seven chapters of the *Third Part of the Biographies of Mathematicians and Astronomers* (疇人傳三編, *Chóu rénzhuàn sān biān*) (on 74 people, adding notes on three other people, referring to 51 others). In the 24th year of the Guāng Xù reign (1898 AD), Huáng Zhōngjùn (黃鍾駿) wrote the 11 chapters of the *Fourth Part of the Biographies of Mathematicians and Astronomers* (疇人傳四編, *Chóu rénzhuàn sì biān*) (with one appendix, a total of 350 people, with references to 86 people). The *Biographies of Mathematicians and Astronomers* has considerable value as regards the understanding the development of the calendar and mathematics in ancient China.

There is an analysis of the *Biographies of Mathematicians and Astronomers* by van Hée (1926) with comments by Mikami (1928).

8.3 The thorough study of Western and Chinese mathematics

In the Qīng Dynasty, besides many scholars collating the mathematics of ancient China, there were also many mathematicians conducting deeper investigations and research based on Western mathematical knowledge that had previously passed into China, or on ancient Chinese mathematical knowledge. They produced a lot of achievements of merit. Although they might be later than in the West, these achievements were the result of independent thought and research by Chinese mathematicians working in a closed-door environment. Also, from the techniques they used, it is clear that they took a different path from those in the West yet arrived at the same conclusions.

Their results include important work on trigonometric expansions, the theory of equations, series, methods of calculating logarithmic tables, and 'Master Sūn's Problem' (on simultaneous congruences) from ancient China. We now briefly describe the results in these areas in the succeeding sections.

1   On the studies of trigonometric expansions

While the Jesuit missionary Pierre Jartoux (杜德美, Dù Déměi) was in China (1701 AD), according to the *Pearls Recovered from the Red River* (赤水遺珍, *Chì shuǐ yí zhēn*) by Méi Juéchéng (梅穀成) he introduced three of the so-called Gregory's formulae (J. Gregory 1638–1675):

$$\pi = 3 + \frac{3.1^2}{4.3!} + \frac{3.1^2.3^2}{4^2.5!} + \frac{3.1^2.3^2.5^2}{4^3.7!} + \ldots \qquad (8.1)$$

$$r \sin \alpha = a - \frac{a^3}{3!r} + \frac{a^5}{5!r^4} - \frac{a^7}{7!r^6} + \ldots \qquad (8.2)$$

$$r \text{ versine } \alpha = \frac{a^2}{2!r} - \frac{a^4}{4!r^3} + \frac{a^6}{6!r^5} - \ldots \qquad (8.3)$$

where $\alpha = \dfrac{a}{r}$

No indication of the proofs of these formulae was given. The Chinese mathematicians used some sort of geometric series to prove these three formulae. They also introduced other formulae. The first to conduct such study was Míng Āntú (明安圖). Later Dǒng Yòuchéng (董祐誠), Xiàng Míngdá (項名達), and Dài Xǔ (戴煦) carried out further investigations.

Míng Āntú, a Mongolian, worked in the State Observatory for a long time. After more than 30 years of hard research he wrote the *Quick Method for Determining Close Ratios in Circle Division* (割圓密率捷法, *Gē yuán mì lǜ jié fǎ*). After his death his student Chén Jìxīn (陳際新) edited the manuscript and completed a total of four chapters in 1774 AD.

Míng Āntú used the method of 'finding the chord knowing the arc' to proceed step by step. In modern algebraic notation his method is as follows. To find the chord AB in Fig. 8.1 knowing the length of the arc ADB.

Divide the arc ADB into m equal parts. Let the small chord of $1/m$th of the arc be C_m and let the chord AB be C. Míng Āntú worked out a formulae for C in terms of the C_m,

$$C = F(C_m).$$

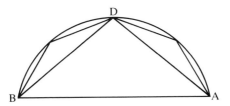

Fig. 8.1 Finding the chord from the arc.

First he calculated for $m = 3, 5, 7, \ldots$ all the odd numbers (that is, three equal parts of the arc, five equal parts, seven equal parts, . . .) and then he calculated for $m = 2, 4, 6, \ldots$, all the even numbers; he also calculated for $m = 10, m = 100, m = 1000, m = 10\,000$. Obviously, when m is very large (for example $m = 10\,000$) mC_m and the arc length are very close. Míng Ăntú then used the approximate value $F(C_{10\,000})$ when $m = 10\,000$ to obtain the final formula for finding the chord, knowing the inscribed polygon (taking mC_m as the arc ABD).

Below we briefly describe the methods of calculation involved when m is odd, even, and 10, 100, 1000, etc. in turn.

When $m = 3$ the formula

$$C = 3C_3 - C_3^3/r^2 \qquad (r \text{ is the radius})$$

can be deduced using the method described in Chapter 16 in the last volume of the *Collected Basic Principles of Mathematics*. Míng Ăntú proceeded to calculate the formulae for $m = 5$. Míng Ăntú's formulae is equivalent to

$$C = 5C_5 - \frac{5C_5^3}{r^2} + \frac{C_5^5}{r^4}.$$

He went on to investigate what happens when $m = 7, 9, \ldots$. From this it follows that when m is odd C can be expressed as the sum of a finite number of terms in C_m.

When m is even it is necessary to expand in an infinite series. Here Míng

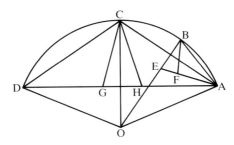

Fig. 8.2 Finding the chord from repeated ratios.

Āntú used a special method repeatedly using geometry and the ratios of line segments.

Here we show Míng Āntú's method of computation for $m = 2$. As in Fig. 8.2, let C be the midpoint of the arc ACD, B the midpoint of the arc ABC, also take AG = DH = AC = DC and let AE = AB.

Then AOB, CAG, GCH, and BAE are similar isosceles triangles, so:

$$OA : AB = AB : BE = AC : GC = GC : GH,$$

from which we get

$$\frac{GH}{BE} = \frac{GC}{AB} = \frac{AC}{OA},$$

so

$$GH = \frac{AC}{OA} \cdot BE .$$

Again from Fig. 8.2, we have

$$AD = AG + DH - GH = 2AC - GH.$$

Substituting the value of GH we then obtain:

$$AD = 2AC - \frac{AC}{OA} \cdot BE,$$

that is,

$$C = 2C_2 - \frac{C_2}{r} \cdot BE$$

(C is the arc length; $r = OA$ = the radius).

Using the same method again, take BE = BF, then the new isosceles triangle is also similar to the previous triangles and using this property we can find the relation between BE, r, C_2, and EF. Continuing in this way and using repeated ratios we finally arrive at

$$C = 2C_2 - \frac{C_2^3}{4r^2} - \frac{C_2^5}{4 \cdot 16 r^4} - \frac{2C_2^7}{4 \cdot 16^2 r^6} - \cdots,$$

which is the formula for $m = 2$.

Again using the formula for $m = 5$ and starting the calculation from two equal half-arcs we easily derive

$$C_2 = 5C_{10} - 5\frac{C_{10}^3}{r^2} + \frac{C_{10}^5}{r^4}.$$

Substituting in the above formula for $m = 2$ we get

$$C = 10C_{10} - 165\frac{C_{10}^3}{4r^2} + 3003\frac{C_{10}^5}{4 \cdot 16 r^4} - \cdots$$

That is to say, the two formulae for $m = 2$ and $m = 5$ give the formula for $m = 10$.

Using the same reasoning and the formula for $m = 10$ one can obtain the formulae for $m = 100$, $m = 1000$, and $m = 10\,000$. The formula calculated by Míng Āntú for $m = 10\,000$ was:

$$C = 10\,000C_{10\,000} - 166\,666\,665\,000\ \frac{C_{10\,000}^3}{4r^2}$$

$$+ 3\,333\,333\,000\,000\,003\,000\ \frac{C_{10\,000}^5}{4 \cdot 16 r^4}$$

$$- 31\,746\,020\,634\,921\,457\,142\,850\,000\ \frac{C_{10\,000}^7}{4 \cdot 16^2 r^6}$$

$$+ \ldots,$$

where here $10\,000C_{10\,000}$ is very close to the arc ACD. Let the arc ACD $= 2a$. Then taking $10\,000C_{10\,000} = 2a$, substituting in the above formula gives:

$$C = 2a - 0.166666665\ \frac{(2a)^3}{4r^2}$$

$$+ 0.03333333\ \frac{(2a)^5}{4 \cdot 16 \cdot r^4}$$

$$- 0.003174602\ \frac{(2a)^7}{4 \cdot 16^2 r^6} + \ldots$$

Here the coefficient of the second term is

$$\frac{0.16666666\ldots}{4} \doteqdot \frac{1}{4 \cdot 3!},$$

the coefficient of the third term is

$$\frac{0.03333333\ldots}{4 \cdot 16} \doteqdot \frac{1}{4^2 \cdot 5!},$$

the coefficient of the fourth term is

$$\frac{0.003174602\ldots}{4 \cdot 16^2} \doteqdot \frac{1}{4^3 \cdot 7!}, \text{ etc.}$$

Substituting in the above formula we get

$$C = 2a - \frac{(2a)^3}{4 \cdot 3! r^2} + \frac{(2a)^5}{4^2 \cdot 5! r^4} - \frac{(2a)^7}{4^3 \cdot 7! r^6} + \ldots \qquad (8.4)$$

Using the method for inverting series one can also calculate

$$2a = C + \frac{1^2 \cdot C^3}{4 \cdot 3! r^2} + \frac{1^2 \cdot 3^2 \cdot C^5}{4^2 \cdot 5! r^4} + \frac{1^2 \cdot 3^2 \cdot 5^2 \cdot C^7}{4^3 \cdot 7! r^6}$$

$$+ \frac{1^2 \cdot 3^2 \cdot 5^2 \cdot 7^2 \cdot C^9}{4^4 \cdot 9! \, r^8} + \cdots \tag{8.5}$$

Let the angle subtended by the arc ACD (that is, $2a$) be α, then from the definitions of the trigonometric functions, $r \sin \alpha = C/2$, and substituting into eqn (8.4) we get:

$$r \sin \alpha = a - \frac{a^3}{3! \, r^2} + \frac{a^5}{5! \, r^4} - \frac{a^7}{7! \, r^6} + \cdots$$

[that is, the Gregory formula eqn (8.2)].
   Again, substituting into eqn (8.5) we get

$$a = r \sin \alpha + \frac{1^2 \cdot (r \sin \alpha)^3}{3! \, r^2} + \frac{1^2 \cdot 3^2 \cdot (r \sin \alpha)^5}{5! \, r^4}$$

$$+ \frac{1^2 \cdot 3^2 \cdot 5^2 \cdot (r \sin \alpha)^7}{7! \, r^6} + \cdots \tag{8.6}$$

Let $\alpha = \pi/6$; then $\sin \alpha = 1/2$, $a = \pi/6$, and substituting into eqn (8.6) and simplifying we get

$$\pi = 3 + \frac{3 \cdot 1^2}{4 \cdot 3!} + \frac{3 \cdot 1^2 \cdot 3^2}{4^2 \cdot 5!} + \frac{3 \cdot 1^2 \cdot 3^2 \cdot 5^2}{4^3 \cdot 7!} + \cdots$$

[this is the Gregory formula (8.1)].
   Míng Āntú also used the repeated ratio method going up to C_{10000}, knowing the arc ACD, to find the sagitta CI as in Fig. 8.3, thereby obtaining:

$$r \cdot \text{vers} \, \alpha = \frac{a^2}{2! \, r} - \frac{a^4}{4! \, r^3} + \frac{a^6}{6! \, r^5} - \frac{a^8}{8! \, r^7} + \cdots$$

[this is the Gregory formula (8.3)].
   Using the inversion method we get

$$a^2 = r \left\{ \frac{2r \cdot \text{vers} \, \alpha}{2!} + \frac{1^2 \cdot (2r \cdot \text{vers} \, \alpha)^2}{4! \, r} \right.$$

$$\left. + \frac{1^2 \cdot 2^2 \cdot (2r \cdot \text{vers} \, \alpha)^3}{6! \, r^2} + \cdots \right\}. \tag{8.7}$$

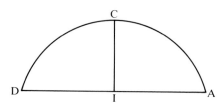

Fig. 8.3 Finding the chord from the sagitta.

Let $b = r$ vers α and at the same time replace a by $2a$ in the above two formulae, then they can be transformed into

$$b = \frac{(2a)^2}{4 \cdot 2! r} - \frac{(2a)^4}{4^2 \cdot 4! 4^3} + \frac{(2a)^6}{4^3 \cdot 6! \cdot r^5} - \dots \qquad (8.8)$$

$$(2a)^2 = r \left\{ 8b + \frac{1^2 \cdot (8b)^2}{4 \cdot 4! r} \right.$$
$$\left. + \frac{1^2 \cdot 2^2 \cdot (8b)^3}{4^2 \cdot 6! r^2} + \dots \right\}. \qquad (8.9)$$

Subsequent to Ming Āntú's introduction of the method of repeated ratios for deriving the above-mentioned formulae, Dǒng Yòuchéng, Xiàng Míngdá, and others continued with further research along these lines.

Dǒng Yòuchéng, whose other name was Fāngli (方立), saw Ming Āntú's work in the year 1819 AD when he was 28, but that was only the manuscript of Chapter 1, which contained nine formulae, so he did not see the complete methods of solution. Dǒng Yòuchéng 'studied it over and over again to find the source of method'; he used a completely different route from Ming Āntú (although he used repeated geometric ratios) and proved these nine formulae.

Dǒng Yòuchéng derived the following four formulae where the chord of the whole arc is c, the sagitta of the whole arc is b, and again the small chord on the nth part of the arc is c, the small sagitta on the nth part of the arc is b:

$$c = nc_n - \frac{n(n^2 - 1)}{4 \cdot 3!} \frac{c_n^3}{r^3}$$
$$+ \frac{n(n^2 - 1^2)(n^2 - 3^2)}{4^2 \cdot 5!} \frac{c_n^5}{r^4} - \dots \qquad (8.10)$$

$$b = n^2 b_n - \frac{n^2(n^2 - 1)}{4!} \cdot \frac{(2b_n)^2}{r}$$
$$+ \frac{n^2(n^2 - 1)(n^2 - 4)}{6!} \cdot \frac{(2b_n)^3}{r^2} - \dots \qquad (8.11)$$

$$c_n = \frac{c}{n} + \frac{(n^2 - 1)}{4 \cdot 3! n^3} \cdot \frac{c^3}{r^2}$$
$$+ \frac{(n^2 - 1)(9n^2 - 1)}{4^2 \cdot 5! n^5} \cdot \frac{c^5}{r^4} + \dots \qquad (8.12)$$

$$b_n = \frac{b}{n^2} + \frac{(n^2 - 1)}{4! n^4} \cdot \frac{(2b)^2}{r}$$
$$+ \frac{(n^2 - 1)(4n^2 - 1)}{6! n^6} \cdot \frac{(2b)^3}{r^2} + \dots \qquad (8.13)$$

but in eqns (8.10) and (8.12) the number n is restricted to odd values.

When $b \to \infty$, $nc_n \to$ the whole arc $2a$ and then the above mentioned four formulae (8.10)–(8.13) can be transformed into Ming Āntú's formulae (8.4), (8.8), (8.5), and (8.9). The other formulae can also be deduced from these.

After he had obtained these results Dǒng Yòuchéng recorded them in his work *Explanation of the Determination of Close Ratios in Circle Division* (using infinite series) (割圜密率圖解 , *Gē yuán mì lü tū jiě*). In 1821, two years after the book was completed, he saw Míng Āntú's completed work, the *Quick Method for Determining Close Ratios in Circle Division* (four chapters), edited by his student. He discovered that although the two men had worked along different lines of thought, nevertheless they had arrived at the same conclusions.

After Dǒng Yòuchéng, Xiàng Míngdá did some further research. He maintained that Dǒng Yòuchéng's results were still incomplete because 'the chord has odd but no even terms, the sagitta has both odd and even appearing'. He went a step further and proved that Dǒng's four formulae can be simplified into two formulae:

$$
c_m = \frac{n}{m} c_n + \frac{n(m^2 - n^2)}{4 \cdot 3! m^3} \cdot \frac{c_n^3}{r^2}
$$

$$
+ \frac{n(m^2 - n^2)(9m^2 - n^2)}{4^2 \cdot 5! m^5} \frac{c_n^5}{r^4} + \ldots \tag{8.14}
$$

$$
b_m = \frac{n^2}{m^2} b_n + \frac{n^2(m^2 - n^2)}{4! m^4} \cdot \frac{(2b_n)^2}{r}
$$

$$
+ \frac{n^2(m^2 - n^2)(4m^2 - n^2)(2b_n)^4}{6! m^6 r^2} + \ldots \tag{8.15}
$$

When $m = 1$ these are Dǒng's two formulae (8.10) and (8.11); when $n = 1$, these are the two formulae (8.12) and (8.13).

The results of Xiàng Míngdá's research were recorded in his work, *The Source of Series* (象數一原 , *Xiàng shú yī yuán*). After his death this book was printed in 1888 AD, edited by Dài Xǔ (戴煦). By that time it was already one and a half centuries after the missionary Pierre Jartoux had brought in the three formulae. At that time the methods of calculus had been used in the West for investigating expansions in series, but this sort of modern method had not reached China. Xiàng Míngdá and the others all worked independently and had nothing to rely on. The patience and perseverance of the Chinese mathematicians in their researches at that time and the results they obtained all deserve praise.

2 Research on the theory of equations

The editing of and research into the mathematical texts of the Sòng and Yuán period enabled the solution of higher degree equations, which seemed to have been lost for several hundred years, to be studied again. The *Collected Basic Principles of Mathematics* and other books contain the method of 'completing the square', which had been introduced from the West. This impelled the scholars into deeper research into equations. The determination of the nature of the roots of equation was investigated with a considerable amount of success. Mathematicians who did research in this area were Wāng Lái and Lǐ Ruì. Wāng Lái (汪萊) (1768–1813 AD), familiar name Xiàoyīng (孝嬰), was also known as Héngzhāi (衡齋). He also wrote a book entitled *Héngzhāi's Mathematics* (衡齋算學, *Héngzhāi suànxué)*; it is a collection of the mathematics written in the course of his life, a total of seven volumes that were printed one at a time (only six volumes were printed while he was alive). After Wāng Lái's death his student Xià Xiè (夏燮) collected and edited his other works and in 1834 AD published the *Unpublished Works of Héngzhāi* (衡齋遺書, *Héngzhāi yí shū)* together with the above *Héngzhāi's Mathematics*. Wāng Lái did a significant amount of work in the area of the theory of equations.

The Sòng and Yuán mathematicians held the view that when they had found one solution of a given problem the problem was completely solved. They had never gone further and investigated whether the equation had another root or what was the nature of the roots, etc. While Wāng Lái was collating and studying the Sòng and Yuán mathematical texts he came across this problem. Through research he knew an equation could have just one positive root or more than one. When an equation had one and only one positive root he maintained that the roots of this equation were completely determined; this situation was called 'fully determined' (確定, Jiāo dìng, lit. '*actually determined*'); when an equation had more than one positive root his view was that the root of the equation could be this one or that one but was still 'not fixed', so this situation was 'not fully determined' (不可知, Bù kě zhī).

Wāng Lái discussed these notions of 'fully determined' and 'not fully determined' in connexion with the interrelations between the coefficients of the equation. This sort of study by Wāng Lái is described in Volume 5 (1810 AD) and Vol. 7 (1810 AD) of *Héngzhāi's Mathematics*.

In Book 5 of *Héngzhāi's Mathematics*, Wāng Lái lists all the various types of quadratic and cubic equation one by one (by the various coefficients being positive or negative and the alternation of signs) and finally he noted whether they were 'fully determined' or 'not fully determined'.

We give several of his problems as examples, namely:

'Given a number of roots plus a number equal to a number of squares: fully determined.'

'Given a number of roots less a number equal to one square: not fully determined.'

'Given a number of roots plus a number equal to a number of cubes: fully determined.'

'Given a number of roots less a number equal to a number of cubes: not fully determined.'

Here 'a number of roots' indicates the linear term, 'a number' is the constant term, 'a number of squares' is the second degree term, 'a number of cubes' is the third degree term and the rest can be inferred similarly. 'Plus' means 'add' but also indicates that the sign of the term's coefficient is positive; 'less' means 'minus', and again this also indicates the sign of the term's coefficient is negative. Thus, written in modern algebraic notation, the examples introduced above have the following forms, where $a, b, c, d > 0$:

an equation of the form $bx + c = ax^2$ has only one positive root (fully determined);

an equation of the form $bx - c = x^2$ has more than one positive root (not fully determined);

an equation of the form $cx + d = ax^3$ has only one positive root (fully determined);

an equation of the form $cx - d = ax^3$ has more than one positive root (not fully determined).

Wāng Lái listed a total of 96 types of quadratic and cubic equations similar.

It is clear that Wāng Lái still had no idea of a general rule. At the same time he tended to maintain that the situation where there is more than one positive root was 'not fully determined' which is, again, not completely accurate. However, the situations discussed by Wāng Lái were new problems that had never been explored by the Sòng and Yuàn mathematicians, and he opened up a new direction in the investigation of the theory of equations during the Qing Dynasty.

Through Jiāo Xún (焦循), Book 5 of *Héngzhāi's Mathematics* by Wāng Lái was made available to Lǐ Ruì. Lǐ Ruì went a step further in investigating the theory by taking Wāng Lái's 96 cases and compressing them into three rules. Lǐ Ruì recorded his three rules in a short essay entitled *A Postscript to Book Five of the Mathematics* [of Héngzhāi] (第五冊算書跋 , *Dì wǔ cè suàn shū bā*). This essay was passed to Wāng Lái through Jiāo Xún's hands. Wāng Lái printed this essay in Volume 6 of *Héngzhāi's Mathematics*.

Lǐ Ruì (1773–1817 AD), familiar name Shàngzhī (尚之), also known as Sìxiāng (四香), was a native of Yuán Hé (元和), the present-day city of Sūzhōu (蘇州). He studied the *Sea Mirror of Circle Measurement, New Steps in Computation*, and several other Sòng and Yuán mathematical texts. He was the personal secretary of Zhāng Dūnrén (張敦仁) and Ruǎn Yuán. He did a lot of work when Zhāng Dūnrén was editing various mathematical works and when Ruǎn Yuán was the chief editor of the *Biographies of Mathematicians and Astronomers*. After his death there appeared the *Collected Mathematical Works of Lǐ Ruì* (李氏算學遺書 , *Lǐ shi suàn xué yí shū*), which included his famous work *The Theory of Extracting Roots* (開方説 , *Kāi fāng shuō*).

Using modern notation Lǐ Ruì's three rules can be written as follows. Suppose the equation is $a_0 x^n + a_1 x^{n-1} + a_2 x^{n-2} + \ldots a_{n-1} x + a_n = 0$, then:

1. If a_0 and a_n have opposite signs and the coefficients of the intermediate terms do not change, i.e. the coefficients of the various terms change sign once only, then the equation has only one positive root (fully determined).

2. If a_0 and a_n have opposite signs, the signs of $a_1, \ldots, a_{n-1}$ are alternately positive and negative and there is a positive root, then the original polynomial can be factorized as:

$$(x - \alpha)(a_0 x^{n-1} + a'_1 x^{n-2} + a'_2 x^{n-3} + \ldots + a'_{n-1}),$$

and if $a'_1, \ldots a'_{n-1}$ have the same sign as a_0, the original equation has only one positive root (fully determined). Otherwise there is more than one positive root (not fully determined).

3. If a_0 and a_n have the same sign then the equation can have more than one positive root (not fully determined).

When Wāng Lái saw Lǐ Ruì's three rules he was extremely impressed by Lǐ Ruì's insight in deducing his rules but he pointed out the incompleteness of the second rule and its impracticality. And in the seventh volume of *Héngzhāi's Mathematics* Wāng Lái presented an alternative method of determining whether an equation has a positive root or not.

Here Wāng Lái suggested an idea that is similar to the present-day so-called 'discriminant' of an equation. For example, concerning determining whether a quadratic equation has a positive root or not, Wāng Lái wrote:

'One square positive, the number of roots negative, the given number positive, take half of the number of roots multiplied by itself and compare it with the given number. When the given number is less than or equal, then there is, if the given number is greater, not.'

This is to say: In a quadratic equation of the form $x^2 - px + q = 0$, when $q \leqslant (p/2)^2$ there is a positive root, when $q > (p/2)$ there is no positive root. $q \leqslant (p/2)^2$ is essentially equivalent to the present-day discriminant of this quadratic equation: $p^2 - 4q \geqslant 0$.

When considering a cubic equation $x^3 - px + q = 0$ Wāng Lái deduced that when $q \leqslant \sqrt{(p/3)}.(2p/3)$ there is a positive root, otherwise there is no positive root. This too is equivalent to the modern-day discriminant of the cubic $4p^3 - 27q^2 \geqslant 0$.

Wāng Lái also started an exploration of problems on the discriminant of higher degree equations (restricted to three-term equations of the form $x^n - px^m + q = 0$).

Whether Volume 7 of Wāng Lái's *Mathematics* passed into Lǐ Ruì's hands is not known. Many years later, in the work *Theory of Extracting Roots* (three chapters), written by Lǐ Ruì, he proposed a more complete method of determining whether an equation has a positive root. His original text says:

If upper negative, lower positive: possible to extract one root. . . .
Upper negative, middle positive, lower negative: possible to extract two roots. . . .

Upper negative, then positive, then negative, lower positive: possible to extract three roots [or one root]. . . .

Upper negative, then positive, then negative, then positive, lower negative: possible to extract four roots [or two roots]. . . .

After this passage there is a passage of commentary which is very helpful for understanding the original passage above. The commentary says:

Suppose we have five terms (that is, a quartic equation), the first two coefficients are negative, the last three positive, then that is 'upper negative, lower positive', that is not saying only the first coefficient is negative and the one last coefficient is positive. The other [cases] are similar.

The 'upper negative, lower positive' indicates that the coefficients in the equation change sign once and this does not depend on the degree of the equation. By the same reasoning 'upper negative, middle positive, lower negative' indicates that the coefficients in the equation have changed sign twice and the rest can be inferred similarly.

Thus, Lǐ Ruì's method of determination is equivalent to:

1. When the coefficients in an equation change sign once it is possible to have one positive root;
2. Two changes of sign, two positive roots;
3. Three changes of sign, three or one positive roots;
4. Four changes of sign, four or two positive roots.

It can be seen that this is the same as what is commonly known nowadays as 'Descartes's rule of signs' (1637 AD).

Lǐ Ruì's *Theory of Extracting Roots* also treats the situation where there are complex roots. He said: 'Not possible to extract roots is called 'no number'. If in a 'no number' case there are two [roots], there is no 'no number' by itself.' 'Not possible to extract roots' indicates the case of complex roots. He considered the complex roots and called them 'no number'. 'If in a "no number" case there are two, there is no "no number" by itself' means that complex roots must occur in pairs and one never appears on its own. This sort of conclusion is quite accurate.

Lǐ Ruì's discussion in the *Theory of Extracting Roots* can be regarded as an admirable conclusion to the persistent research of the two men Wāng Lái and Lǐ Ruì. Although the results of Wāng Lái and Lǐ Ruì's researches are somewhat later than the results in Western mathematics, nevertheless they obtained the results independently and had nothing to start from. They were able to break the confines of Sòng and Yuán mathematics, progressing a step further, stimulating each other and researching: this sort of spirit is praiseworthy.

3   On the investigations into summing finite series

In the middle of the Qīng period mathematicians investigated aspects of the sums of finite series. People who undertook such research were Chén Shìrén (陳世仁), Wāng Lái, Dǒng Yòuchéng, Luó Shìlín and Lǐ Shànlán (李善蘭). Amongst these the most outstanding research was done by Chén Shìrén and Lǐ Shànlán. Luó Shìlín's work was limited to commenting on Zhū Shìjié's *Precious Mirror of the Four Elements*, nor did Wāng Lái or Dǒng Yòuchéng's work surpass the achievements of the Sòng and Yuán mathematicians. However, Chén Shìrén and Lǐ Shànlán's work was different. Here we briefly describe Chén Shìrén and Lǐ Shànlán's investigations in this area were different.

Chén Shìrén (1676–1722 AD) lived during the Kāng Xī reign. He wrote the *Supplement to 'What Width'* (少廣補遺, *Shǎo guǎng bǔ yì*), one chapter, which was compiled into the *Complete Library*. In the *Supplement to 'What Width'* Chén Shìrén described seven systems (the first can be divided into two so in fact there are really eight systems) with a total of 37 formulae for the sums of finite series. Here for the sake of convenience we use modern mathematical notation to illustrate these formulae. Chén Shìrén's were based on the following two systems of formulae:

$$\sum_{r=1}^{n} r \qquad \ldots \text{Chén Shìrén called this 'halving piles'}$$

$$\sum_{r=1}^{n} \frac{r(r+1)}{2} \qquad \ldots \text{'standing piles'}$$

$$\sum_{r=1}^{n} 2^{r-1} \qquad \ldots \text{'doubling piles'}$$

$$\sum_{r=1}^{n} r^2 \qquad \ldots \text{'square piles'}$$

$$\sum_{r=1}^{n} r^3 \qquad \ldots \text{'cubic piles'}$$

(8.16)

$$\sum_{r=m}^{n} r \qquad \ldots \text{'halving pile, partial sum'}$$

$$\sum_{r=m}^{n} \frac{r(r+1)}{2} \qquad \ldots \text{'standing pile, partial sum'}$$

$$\sum_{r=m}^{n} r^2 \qquad \ldots \text{'square pile, partial sum'.}$$

(8.17)

System (8.16) starts calculating from $r = 1$, that is, starting from the top of the pile. System (8.17) starts calculating from $r = m$, that is, equivalent to starting calculating part-way, that is why it is called a 'partial sum'.

Most of the formulae for these two systems had already been found by the Sòng and Yuán mathematicians, though those mathematicians had not given actual formulae for the general cases. The special feature of Chén Shírén's investigations is that after determining the various formulae in the systems (8.16) and (8.17) he omitted the odd terms (or the even terms) and then found the respective sums. The formulae for finding the various sums after omitting the even terms were very fruitful. For example, in 'halving piles': omitting the even terms from $1 + 2 + 3 + 4 + \ldots + n$ one obtains

'omitting evens, halving pile': $1 + 3 + 5 + \ldots + \overline{2n - 1} = n^2$,

'omitting evens, standing pile': $1 + (1 + 3) + (1 + 3 + 5) +$

$$+ (1 + 3 + 5 + \ldots + \overline{2n - 1}) = \frac{n}{3}\left(n^2 + \frac{3}{2}n + \frac{1}{2}\right),$$

'omitting evens, square pile': $1^2 + 3^2 + 5^2 + \ldots + \overline{2n - 1}^2$

$$= \frac{n}{3}(4n^2 - 1),$$

'omitting evens, cubic pile': $1^3 + 3^3 + 5^3 + \ldots + \overline{2n - 1}^3$

$$= n^2(2n^2 - 1), \text{ etc.}$$

Similarly various systems of formulae can be derived in the 'standing piles' case and 'partial sums' by 'omitting the odds' or 'omitting the evens' from the 'halving piles', 'standing piles', 'cubic piles', etc.

The Sòng and Yuán mathematicians did not discuss 'omitting odds' or 'omitting evens' and then finding the sums. Neither did the Western mathematics that came in before the Kāng Xī (康熙) reign (1662–1722) mention knowledge of this area. Unfortunately Chén Shírén's manuscript was not published and circulated so there were not many people who understood its contents.

After Chén Shírén, although there were many mathematicians who researched into the problem of summing finite series, the most outstanding success must be the results of Lǐ Shànlán 150 years later, as has been mentioned above.

Lǐ Shànlán (李善蘭) (1811–1882 AD), familiar name Rénshū (壬叔), also known as Qiūrèn (秋紉), was a native of Hàiníng (海寧) in Zhèjiāng (浙江) province. His investigations on series are presented in his work *Sums of Piles of Various Types* (垛積比類 , *Duò jī bǐ lèi*). The complete book of *Sums of Piles of Various Types* consists of four chapters and each chapter forms a complete system. In outline the four chapters are structured about the following four systems:

1. Chapter 1, 'triangular piles'. This can be divided into several subsystems. These systems can be summarily indicated using modern mathe-

matical notation, by the following formulae:

    (a) 'Triangular pile'

$$\sum \frac{1}{p!} r(r+1)(r+2)\ldots(r+\overline{p-2})(r+\overline{p-1})$$

$$= \frac{1}{(p+1)!} n(n+1)(n+2)\ldots(n+\overline{p-1})(n+p), \tag{8.18}$$

    (b) '(Triangular pile) multiplying by supporting pile'

$$\sum \frac{1}{p!} r(r+1)(r+2)\ldots(r+\overline{p-2})(2r+\overline{p-2})$$

$$= \frac{1}{(p+1)!} n(n+1)(n+2)\ldots(n+\overline{p-1})(2n+\overline{p-1}), \tag{8.19}$$

    (c) '(Triangular pile) double multiplying by supporting pile'

$$\sum \frac{1}{p!} r(r+1)(r+2)\ldots(r+\overline{p-2})(3r+\overline{p-3})$$

$$= \frac{1}{(p+1)!} n(n+1)(n+2)\ldots(n+\overline{p-1})(3n+\overline{p-2}), \tag{8.21}$$

    (d) '(Triangular pile) triply multiplying by supporting pile'

$$\sum \frac{1}{p!} r(r+1)(r+2)\ldots(r+\overline{p-2})(4r+\overline{p-4})$$

$$= \frac{1}{(p+1)!} n(n+1)(n+2)\ldots(n+\overline{p-1})(4n+\overline{p-3}). \tag{8.22}$$

Taking $p = 1, 2, 3, 4, \ldots$ one can obtain the formulae for each of the subsystems. It is worth pointing out that the formulae in these four systems all possess a common feature, namely that in each system the result of the preceding formula is always the general term of the following formula. For example, the result when $p = 1$ is just the general term of the series for $p = 2$; the result when $p = 2$ is the general term of the series for $p = 3$; and so on.

In fact the formulae for these four subsystems can be gathered into one formula as

$$\sum \frac{1}{p!} r(r+1)(r+2)\ldots(r+\overline{p-2})(mr+\overline{p-m})$$

$$= \frac{1}{(p-1)!} n(n+1)(n+2)\ldots(n+\overline{p-1})(mr+\overline{p-m+1}).$$

For $m = 1, 2, 3, 4$ this is the above-noted formulae (8.18)–(8.22), respectively.

    2. Chapter II, 'square piles'. This can be divided into several subsystems.

(a) 'Square piles'. This indicates the problems of finding the sums of series of the form Σr^p. Lǐ Shànlán's method of solution is equivalent to the following formula:

$$\Sigma r^p = \Sigma \left[A_p^i \overset{n-i+1}{\Sigma} f_p^r \right],$$

where $f_p^r = \dfrac{1}{p!} r(r+1)(r+2) \ldots (r+\overline{p-1})$ and A_p^i is the ith

coefficient in Σr^p. Lǐ Shànlán obtained these coefficients from the following table:

1	Coefficient for $p = 1$
1, 1	Coefficients for $p = 2$
1, 4, 1	Coefficients for $p = 3$
1, 11, 11, 1	Coefficients for $p = 4$
1, 26, 66, 26, 1	Coefficients for $p = 5$
. . .	. . .

On the method of computing this table, Lǐ Shànlán said: 'Each entry depends on the entries immediately above to left and right, the left entry governs the number of layers along the left diagonal and the right entry governs the following layers along the right diagonal and each multiplies it according to the layer number. Combining them gives the present entry.' The A_p^i can be determined from the following rule (using modern mathematical notation):

$$\begin{cases} \text{When } i = 1 \text{ or } i = p, \ A_p^i = 1; \\ \text{When } 1 < i < p, \ A_p^i = (p - i + 1)A_{p-1}^{i-1} + i\,A_{p-1}^i. \end{cases}$$

The general formulae for Lǐ Shànlán's 'square piles' are:

'basic pile' (when $p = 1$):

$$\Sigma r = \overset{n}{\Sigma} f_1^r,$$

'square pile' ($p = 2$):

$$\Sigma r^2 = \overset{n}{\Sigma} f_2^r + \overset{n-1}{\Sigma} f_2^r,$$

'cubic pile' ($p = 3$):

$$\Sigma r^3 = \overset{n}{\Sigma} f_3^r + 4 \overset{n-1}{\Sigma} f_3^r + \overset{n-2}{\Sigma} f_3^r,$$

'quartic pile' ($p = 4$):

$$\Sigma r^4 = \overset{n}{\Sigma} f_4^r + 11 \overset{n-1}{\Sigma} f_4^r + 11 \overset{n-2}{\Sigma} f_4^r + \overset{n-3}{\Sigma} f_4^r,$$

That is to say, the problem of finding the sum of a series of the form Σr^p can be reduced to the sums of p-multiplying triangular piles. These are calculated one by one from the first layer, second layer, third layer, ... Each is multiplied by a number from a sequence computed in a separate table and then they are added together to give the sum.

(b) 'Squaring square supporting pile'. This is similar to the 'square supporting pile' in System 1. In fact it is the problem of finding the sums of Σr^2, $\Sigma[\Sigma r^2]$, $\Sigma\{\Sigma[\Sigma r^2]\}$, . . . (Below we use the notation Σ^p to stand for this pth iterated sum.)

(c) 'Cubic square supporting pile'. This can be obtained using the following formula:

$$\Sigma p_r^3 = \sum^{n} f_{p+2}^r + 4 \sum^{n-1} f_{p+2}^r + \sum^{n-2} f_{p+2}^r.$$

Letting $p = 1, 2, 3, 4, \ldots$ yields all the formulae.

(d) 'Quartic square supporting pile'. In fact these are problems of the form Σp_r^4 and the method of solution is similar to (c).

3. Chapter III, 'triangular self-multiplying piles'.

(a) Triangular self-multiplying pile'. This is the problem of finding the sums of the squares of each of the terms (f_p^r) for the triangular pile. They can be summarized by the following formula:

$$\Sigma (f_p^r)^2 = \sum_{i=1}^{p+1} \left[(A_p^i)^2 \sum_{r=1}^{n-i+1} f_{2p}^r \right].$$

Here A_p^i denotes the coefficient of the ith term in the expansion of the binomial $(a + b)^p$. This is the internationally known 'Lǐ Shànlán's identity'. Taking $p = 1, 2, 3, 4, \ldots$ yields in succession the 'A pile', 'B pile', 'C pile', etc. in Lǐ Shànlán's book. For example,

'A pile' $(p = 1)$:

$$\Sigma (f_1^r)^2 = 1^2 \sum_{r=1}^{n} f_2^r + 1^2 \sum_{r=1}^{n-1} f_2^r,$$

'B pile' $(p = 2)$:

$$\Sigma (f_2^r)^2 = 1^2 \sum_{r=1}^{n} f_4^r + 2^2 \sum_{r=1}^{n-1} f_4^r + 1^2 \sum_{r=1}^{n-1} f_4^r,$$

'C pile' $(p = 3)$:

$$\Sigma(f_3^r)^2 = 1^2 \sum_{r=1}^{n} f_6^r + 3^2 \sum_{r=1}^{n-1} f_6^r$$

$$+ 3^2 \sum_{r=1}^{n-2} f_6^r + 1^2 \sum_{r=1}^{n-3} f_6^r,$$

. . .

From these it can be seen that when Lǐ Shànlán treated problems of the form $\Sigma(f_p^r)^2$ he still converted them into $2p$-multiplying triangular piles, the first starting from the first layer, the second from the second, and so on, each term then being multiplied by a previously determined coefficient.

(b) In Chapter III there are further 'supporting pile' problems similar to those in the previous two chapters. Lǐ Shànlán had also converted them one by one into a series of cubic piles and then found the sums. We shall not discuss these because of limitations of space.

4. 'Modified triangular piles'. These are problems on finding the sums of forms such as $\Sigma r \cdot f_p^r$, $\Sigma r^2 \cdot f_p^r$, $\Sigma r^3 \cdot f_p^2$. These are similar to the 'peak-shaped piles' in Zhū Shìjié of the Yuán dynasty's piling problems (see Chapter 5, p. 156 above). Lǐ Shànlán's method of solution is again to convert them into triangular piles, then multiply by predetermined coefficients and then obtain the sum. For example:

(a) 'Modified triangular pile' can be constructed as the following formulae:

$$\sum r \cdot f_p^r = \sum_{r=1}^{n} f_{p+1}^r + p \sum_{r=1}^{n-1} f_{p+1}^r,$$

$$(p = 1, 2, 3, \ldots).$$

(b) 'second modified triangular pile':

$$\sum r^2 \cdot f_p^r = \sum_{r=1}^{n} f_{p+2}^r + (1 + 3p) \sum_{r=1}^{n-1} f_{p+2}^r$$

$$+ p^2 \sum_{r=1}^{n-2} f_{p+2}^r, (p = 1, 2, 3, \ldots).$$

(c) 'third modified triangular pile':

$$\sum r^3 \cdot f_p^r = \sum_{r=1}^{n} f_{p+3}^r + (4 + 7p) \sum_{r=1}^{n-1} f_{p+3}^r$$

$$+ [(2p + 1)^2 + 2p^2] \sum_{r=1}^{n-2} f_{p+3}^r + p^3 \sum_{r=1}^{n-3} f_{p+3}^r$$

$$(p = 1, 2, 3, \ldots).$$

The above is a general description of Lǐ Shànlán's investigations into the problem of sums of finite series. Lǐ Shànlán also combined the research into the above system of formulae and the 'source of the method of extracting roots' (that is, the so-called Pascal's triangle) and gave his research a novel feature which had 'tables, diagrams, and method'.

It is worth pointing out that Lǐ Shànlán only listed these identities, he did not give complete proofs. Because the original text is so terse it is difficult to infer Lǐ Shànlán's complete line of thought; this awaits further investigation.

4 Research in other areas

In addition to the fine results in the expansion of trigonometric functions, the theory of equations and the summation of finite series described above, the mathematicians in the middle of the Qīng also achieved some good results in several other areas such as number theory, the study of ellipses, the binomial theorem and the expansion of logarithmic functions in infinite series. We shall now briefly present achievements in these areas. The reason these are presented together is not to suppose they have any relation to one another, nor is it to say that they are not as important as what has already been described; it is because they are all isolated and relatively difficult that they are treated together so that something may be said briefly about each.

In the area of number theory the important work concerns the further researches into the 'technique of finding 1 by the great extension'. There were a lot of works in this area, the main ones being Zhāng Dūnrén's *Technique of Finding 1* (*Indeterminate Analysis*, 求一算術, *Qiú yī suàn shù*, 1831 AD), Jiāo Xún's (焦循) *Technique of finding 1 by the great extension* (大衍求一術, *Dà yǎn qiú yī shù*), Luò Téngfèng's (駱騰鳳) *Technique of finding 1 by the great extension* (大衍求一法, *Dà yǎn qiú yī fǎ*) in his *Records of the Art of Learning* (藝游錄, *Yì yóu lù*) (1815 AD), *Shí Rìchún's* (時日淳) *Path to the Technique of Finding 1* (求一術指, *Qiú yī shù zhǐ*, 1873 AD) and Huáng Zóngxiàn's (黃宗憲) *Explanation of the Technique of finding 1* (求一術通解, *Qiú yī shù tōng jiě*, 1874 AD). In particular, Huáng Zóngxiàn introduced the concept of 'root numbers' (數根, shù gēn) that is, prime numbers, and prime factor decomposition, which permitted a clearer understanding of Qín Jiǔsháo's 'technique of finding 1 by the great extension'.

In the area of number theory we should also mention Lǐ Shànlán's research in this area in his paper *Four Techniques of Determining Primes* (考數根四法, *Kǎo shù gēn sì fǎ*) which was first published in 1872. Borrowing modern mathematical notation to write them Lǐ Shànlán's conclusions are: if $a^d - 1$ is divisible by N and N is a prime number then $N - 1$ is divisible by d. However, $N - 1$ being divisible by d is only a necessary condition for N to be a prime number and not a sufficient condition. Thus Lǐ Shànlán had proved the famous Fermat Little Theorem (Fermat, French mathematician, 1601–1665 AD) and had pointed out at the same time that the converse of the theorem is not true. Although this result was obtained much later than in Europe, it was derived independently by Lǐ Shànlán in a situation isolated from Western influence. The *Four Techniques of Determining Primes* also contains the results of other research, which we shall not list here item by item.

In the developments in the area of finite series, besides the investigations by Míng Āntú, Dǒng Yòuchéng, and Xiàng Míngdá in the area of the expansion of trigonometric functions noted in the previous section, there were other explorations into the expansions of various functions by Xiàng Míngdá, Dài Xǔ, and Lǐ Shànlán which are worth mentioning.

Xiàng Míngdá used his own unique method to find the length of the circumference of an ellipse.

In the year 1846 Dài Xǔ proved the binomial theorem. His result presented in modern mathematical notation is: the binomial expansion of

$$(1 + \alpha)^m = 1 + m\alpha + \frac{m(m - 1)}{2!}\alpha^2$$

$$+ \frac{m(m - 1)(m - 2)}{3!}\alpha^3 + \ldots,$$

which is established for any rational number m when $|\alpha| < 1$. This conclusion was obtained by Newton in Europe as early as 1676 AD but it was not until the year 1859 AD that it was introduced into China in the book *Algebra*, translated by Lǐ Shànlán. Dài Xǔ's research had been obtained independently before the relevant Western mathematical knowledge came in.

It is also worth selecting some important points from the investigations with the technique of 'cones' by Lǐ Shànlán, Dài Xǔ's work on logarithms and Lǐ and Dài's work in the area of the expansion of trigonometric functions in order to introduce them.

In the three books *Explanation of the Square and the Circle (方圓闡幽, Fāng yuán chǎn yōu)*, *Unveiling the Secrets of Arc and Sagitta (弧矢啓秘, Hú shǐ qǐ mì)* and *Seeking the Source of Logarithms (對數探源, Duì shù tàn yuán)* (completed before 1846 AD), Lǐ Shànlán used his unique technique of 'circular cones' to bring out a lot of significant results.

At the very beginning of the *Explanation of the Square and the Circle*, ten preparatory theorems are given. The central ideas can be summarized in modern mathematical notation as follows: when $n \geqslant 2$, x^n can be represented by a plane area or a line segment. Take a conical body whose cross-sectional area at height h is always equal to ah^n. Then the volume of such a conical body (height h, base area ah^n) is equal to $ah^n/(n + 1)$. This is equivalent to giving the definite integral formula:

$$\int_0^h ax^n dx = \frac{ah^{n+1}}{n + 1}.$$

He also derived the conclusion that such conical bodies of the same height can be combined into a cone. This is equivalent to deriving the formula:

$$\int_0^h a_1 x dx + \int_0^h a_2 x^2 dx + \int_0^h a_3 x^3 dx + \ldots + \int_0^h a_n x^n dx$$

$$= \int_0^h (a_1x + a_2x^2 + \ldots + a_nx^n)dx.$$

Lǐ Shànlán used this type of technique of 'cones' to calculate the area of a circle. As in Fig. 8.4, divide the cone BAC into many cones BAD, DAE, EAF, FAG, . . . of the same height. Let the radius of the circle be 1, then using the above preparatory theorems the total area (A) of the cone BAC is finally calculated as:

$$A = \frac{1}{2} \cdot \frac{1}{3} + \frac{1}{8} \cdot \frac{1}{5} + \frac{1}{16} \cdot \frac{1}{7} + \frac{5}{128} \cdot \frac{1}{9} + \ldots,$$

from which the area of the circle radius $r = 1$ can be shown to be equal to:

$$\pi r^2 = \pi = 4 - 4A = 4 - 4\left(\frac{1}{2} \cdot \frac{1}{3} + \frac{1}{8} \cdot \frac{1}{5} + \frac{1}{16} \cdot \frac{1}{7} + \frac{5}{128} \cdot \frac{1}{9} + \ldots\right).$$

In his book *Seeking the Source of Logarithms* Lǐ Shànlán also used this type of technique of 'cones' to obtain a formula for the natural logarithm, $\log_e n$, of a number n as:

$$\log_e n = \left(\frac{n-1}{n}\right) + \frac{1}{2}\left(\frac{n-1}{n}\right)^2 + \frac{1}{3}\left(\frac{n-1}{n}\right)^3$$
$$+ \frac{1}{4}\left(\frac{n-1}{n}\right)^4 + \ldots.$$

Dài Xǔ used another method in his *Concise Technique of Logarithms* (對數簡法 , *Duì shù jiān fǎ*, 1846 AD), where he also derived a formula for $\log_e n$. Using this formula to compute logarithmic tables is much simpler and faster than the method used in the *Collected Basic Principles of Mathematics*.

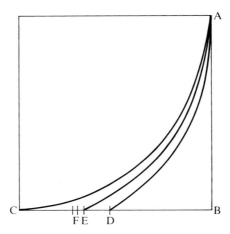

Fig. 8.4 Lǐ's diagram.

Further, in his book *Unveiling the Secrets of Arc and Sagitta*, Lǐ Shànlán again used his 'technique of cones' to derive the formula for the tangent given the arc. This formula is equivalent to

$$\tan a = a + \frac{a^3}{3} + \frac{2a^5}{15} + \frac{17a^7}{315} + \cdots$$

Dài Xǔ also derived the same result as Lǐ Shànlán using Míng Āntu's method of 'finding the chord knowing the arc' (see above, Chapter 8, p. 234).

Although most of the above-mentioned results were derived by the Chinese mathematicians later than in the West, they were, however, all achieved in an environment where this type of mathematical knowledge had not been passed in from the West and they were obtained through their own independent research. It is clearly evident that beginning with Míng Āntu through Dài Xǔ and Lǐ Shànlán, a lot of the work by the Chinese mathematicians of that time was approaching the fundamental concepts of differentiation and integration. Some non-Chinese have also accepted this point. For example, in the preface of the joint translation by Alexander Wylie and Lǐ Shànlán of a *Manual on Calculus* (代微積拾級 , *Dài wēi jī shí jī*) it says: '. . . differentiation and integration are not found in mathematics books in China, but when one studies the astronomers and mathematicians of that time such as Dǒng Yòuchéng, Xiàng Míngdá, Xú Yǒurén, Dài Xǔ, Gù Guānguāng and Lǐ Shànlán and their works, their reasoning is very close to infinitesimals . . .'. After Joseph Edkins, who jointly translated *Mechanics* (重學 , *Zhòng xué*) with Lǐ Shànlán, saw Dài Xǔ's work, he was greatly impressed, did the translation, and sent it to the Mathematical Society of his own country. This might be the earliest research work by a Chinese mathematician translated into a foreign language.

From the above it is clear that if the Western techniques of differentiation and integration had not passed into China, the Chinese mathematicians could also have progressed by their own way from elementary mathematics along the path of higher mathematics.

9
The second entry of Western mathematics into China

9.1 A brief description of the entry of Western mathematics for the second time

In 1840 AD the Opium Wars broke out, and after the defeat of China the Qīng government's closed-door policy collapsed. The Western powers increased their encroachment on China. A number of unequal treaties were signed in succession. The 'closed door' was broken down by Western guns and cannon. From then on, China set forth on the path to becoming a half feudal, half colonial society. Further, a history of almost 100 years of struggle by the Chinese against imperialism began.

After the signing of the Tiānjīn (天津, Tientsin) Treaty and the Běijīng Treaty in 1858–1860 AD, the 'foreign affairs group' (洋務派, Yáng wù pài) gained important political influence. At that time a lot of people simplistically put the basis for China being invaded on the foreigners' 'strong ships and superior arsenal'. The Chinese had no steamboats and no cannon and no way to defend themselves against the Westerners. So they had the view that in order to strengthen China they had to start learning from the West. So 'foreign affairs' (Yáng-wù) and 'Western learning' (西學, Xī xué) prevailed in those days. Institutes such as the 'Language Institute' (同文館, Tóng wén guǎn) and the 'South of the River Manufacturing Company' (江南製造局, Jiāng nán zhǐ zào jú) and so on were established. The previous situation whereby little of Western mathematics had been translated ceased and the work of translating Western mathematics rapidly developed.

In the 1850s and 1860s a set of books was translated by the famous mathematician Lǐ Shànlán. In the 1870s another set of works was translated by Huà Héngfāng (華蘅芳).

In fact the period after the Opium War was the second time Western mathematics came into China, or the stage of re-entry (as opposed to the first entry before the Kāng Xī reign). There was a lot of difference in the content of the mathematical knowledge passed in during the two distinct stages. The second time the material was such things as analytic geometry, differential and integral calculus, probability theory, etc. and all of these were outside the confines of elementary mathematics (that is, mathematics without variables), but within higher mathematics (that is the mathematics of variables). The situations were very different for the start of these two entries. At the second entry the Western mathematics was accompanied by a revolution in the educational system and a change in textbooks; its sphere of influence became

255

wider and wider. These three combined, that is, the re-entry of Western mathematics, the revolution in the educational system, and the translation of textbooks, worked together to establish the main content of the mathematical history of this period. By the beginning of the 20th century the old civil service examination system had finally been abolished (in 1905) and various kinds of Western school had begun to be established. The textbooks used in the various kinds of Western school were changed again and again until finally there was essentially no difference between them and the mathematical textbooks used all over the world. The other types of mathematical methods from ancient China, as well as abacus calculation (which was still in wide use throughout the populace) had already blended with the commonly used international mathematical methods, including the methods of recording numbers and mathematical symbols.

The advent of Western mathematical knowledge into China through the three stages of 'entry, assimilation, and re-entry' altogether took more than 250 years.

By the 1920s Chinese mathematicians had already begun to produce results in several areas of modern mathematics. Since then the story of Chinese mathematics has again embarked on a new era in history.

9.2 The translations of Western mathematical books

1   The translation by Lǐ Shànlán (李善蘭)

Lǐ Shànlán (Fig. 9.1) was a famous Chinese mathematician of the 19th century. Before he had contact with the higher Western mathematics such as calculus, etc., he had already done some creative research in several areas of mathematics (see Sections 8.3.3 and 8.3.4). By chance he lived during the time when Western mathematics entered China for the second time and consequently he was also an important figure in introducing new mathematical knowledge from the West. In trying to get an overall assessment of Lǐ Shànlán throughout the whole of his mathematical life it is difficult to decide which occupied the more important position: his translations of Western mathematics or his creative research. In fact he made important contributions in both these areas.

His creative research has been briefly described in the previous sections. In this section we should like to introduce some of the contributions he made in the work of translation.

When Lǐ Shànlán was 10 years old he started studying the *Nine Chapters on the Mathematical Art* and when he was 15 he read *Euclid's Elements* (the translation of the first six books by Xú Guāngqǐ). After he encountered the Sòng and Yuán mathematical texts in the Hángzhōu (杭州) civil service examination (such as Lǐ Zhì's *Sea Mirror*) his interest in mathematics grew. At the same time he knew a lot of the mathematicians of that time personally.

Subsequently he himself achieved outstanding results in several areas of mathematics. In the year 1852 AD he arrived in Shànghǎi and got to know the Englishman Alexander Wylie. From then on he and Alexander Wylie translated books together. The main books they translated were the following:

1. *Euclid's Elements* (the last nine books). They completed the translation in the four years from 1852 to 1856 AD, including two books dubiously attributed to Euclid. It was two and a half centuries after the translation of the first six books by Xú Guāngqǐ and others (1606 AD) that the complete work by Euclid was translated into Chinese. However, the book that Lǐ Shànlán and Wylie worked from was not in Greek but was an English translation of the period. The Chinese translation, at last completed by Lǐ Shànlán and which is still in circulation, remains up to the present the only Chinese translation.

2. The *Elements of Algebra* by A. De Morgan (1835 AD), translated in 1859 by Lǐ Shànlán and A. Wylie as 代數學, *Dài shù xué*. This was the first translation into Chinese of a modern Western algebra. Lǐ Shànlán maintained that the special features of this branch of mathematics were: 'using letters to represent numbers: either variables or unknown constants. . . . letters are also substituted for frequently used constants which are complicated. So this field of mathematics, which has the name 'algebra', he translated appropriately as the 'knowledge of substituting numbers' (代數學, Dài shù xué). This is the first time the name 'knowledge of substituting numbers' appeared. This name is still used nowadays and this translation of the name was also adopted in Japan.

3. *Eighteen Chapters of Elements of Analytical Geometry and of Differential and Integral Calculus* by E. Loomis (1850 AD), translated as (代微積拾級 , *Dài wéi jī shí jī*) (the translation was completed in 1859 AD). This was the first translation into Chinese of an Analytical Geometry and Calculus which entered China. This book was used as a textbook in analytical geometry and calculus for the next several decades. It says in the preface to this book that it was written in Leibniz's notation. But in fact it was changed into a Chinese form: the notation used for differentiation is the radical ' 彳 (rén)' (see Appendix), which is the radical of the character 微, wēi (literally, infinitesimal); the notation for integration is 禾 (hé) (the radical of 積 jī, integrate). For example $\int 3x^2 dx$ was represented by 禾三天 = 彳天 , hé sān tiān = rén tiān where 天, tiān, meaning heaven, is used as an unknown, x, (cf. Zhū Shìjié above).

In addition to the above three books, Lǐ Shànlán translated the *Outlines of Astronomy* by Herschel (1849) (as 談天 , *Tán tiān*) and several chapters of Newton's *Principia* in conjunction with Alexander Wylie.

Lǐ Shànlán also translated the *Theory of Conic Sections*, three chapters, in conjunction with Joseph Edkins, the English missionary, completing the translation in 1866. This book treats various properties of conics. Lǐ Shànlán and Joseph Edkins also cooperated in translating *An Elementary Treatise on Mechanics* by William Whewell. Summing up, Lǐ Shànlán's translation work was a contribution to introducing the relatively new Western knowledge of mathematics. This new knowledge formed part of the most important content of mathematics entering for the second time from the West. The books he translated were several of the earliest higher mathematics works passed into China.

Lǐ Shànlán's translation work was very careful. In the preface to the last nine books of *Euclid's Elements* Alexander Wylie said: 'I (Alexander Wylie) translated orally, he (Lǐ Shànlán) wrote it down, cutting out the superfluous and attending to the accuracy, examining it again and again so as to make it perfect, that was essentially his work . . .'. It is evident that Lǐ Shànlán put a great deal of work into the translation of *Euclid's Elements*.

There is another point that should be mentioned. The works translated by Lǐ Shànlán contained very new material, and contained much that was too difficult for the ordinary reader at that time. The circulation and influence of these translations was confined to a small number of lovers of mathematics. Consequently later on people criticized his translations, saying his aim was to 'show off his specialist learning'. Although this sort of comment was not fair, it is not entirely without foundation.

Fig. 9.1 Lǐ Shànlán.

2 Huà Héngfāng's translations

Lǐ Shànlán translated a lot of mathematics books from the West that were from the 1850s. About 20 years later Huà Héngfāng and others also translated a set of mathematical texts. Huà Héngfāng (Fig. 9.2) was a mathematician after Lǐ Shànlán who made very large contributions to the introduction of Western mathematics.

Huà Héngfāng (華蘅芳, 1833–1902 AD), familiar name Ruò Tīng (若汀), was a native of Jīn Kuì (金匱) in Jiāngsū (江蘇) province [at the present time a district in the principality of Wú xī (無錫)]. At the age of 14 he understood Chéng Dàwèi's *Systematic Treatise on Arithmetic* (算法統宗, *Suàn fǎ tǒng zóng*). Later he worked very hard on the *Nine Chapters on the Mathematical Art*, mathematical texts of the Sòng and Yuán period, and works in the *Collected Basic Principles of Mathematics*. After growing up he became acquainted with Lǐ Shànlán.

In 1868 the Shànghài Manufacturing Company established the Translation Institute and Huà Héngfāng was appointed to the Institute to work on the translation of mathematical and other scientific works. He and the English missionary John Fryer together translated a number of Western mathematical works, including important items from the eighth edition of the *Encyclopaedia Britannica*:

Fig. 9.2 Huà Héngfāng.

1. Works by the Englishman Wallace — *Algebra*, (代數術, *Dài shù xué*) 25 chapters, translated in 1873 AD; *Fluxions* (微積溯源, *Wēijī sùyuán*) eight chapters, translated in 1878 AD.

2. The Englishman, J. Hymers — *A Treatise on Plane and Spherical Trigonometry* (三角數理, *Sān jiǎo shù lǐ*), 12 chapters, translated in 1877 AD.

3. The Englishman T. Lund — *A Companion to Wood's Algebra* (代數難題, *Dài shù nán tí*), 16 chapters, translated in 1883 AD.

4. The Englishman Galloway — *Probability* (from *Encyclopaedia Britannica*) and Anderson: *Probabilities* (from *Chambers Encyclopaedia*) (決疑數學, *Jué yí shù xué*) 10 chapters, translated in 1880 AD.

5. The Englishman Ball — *Combinatorics*. (合數術, *Gě shù shù*) 11 chapters, translated in 1888 AD.

These works systematically introduced the various branches of mathematics: algebra, trigonometry, calculus, and probability. The contents of these various translations were somewhat easier and broader than Lǐ Shànlán's translations and the language of the translations was very fluent. Also, several of these books were adopted by Westernized schools as textbooks.

Among all these translations *Probabilities* is worth particular mention. This book introduced a very new branch of mathematics — probability theory. The preface of this book gives a brief description of the history of the theory of probability from its invention by Laplace in 1812 (Laplace, French, 1749–1827 AD). Besides the general knowledge on probability, this book also introduced the method of least squares and at the same time described various applications of probability to astronomy, surveying, and theoretical physics. This was a part of the mathematics being imported at that time that was relatively difficult, and it was also some of the most recent material in the world history of mathematics.

As well as translating Western texts, Huà Héngfāng also wrote several mathematical papers, which were collected into a volume entitled *Mathematical papers from the Xing Sù Study* (行素軒算稿, *Xíngsù xuān suàn gǎo*).

Most of the content of this book is very simple and there is not much original research in it. Huà Héngfāng and Lǐ Shànlán were the important figures in this period when Western mathematics came in for the second time, but from the point of view of original research Huà Héngfāng was much inferior to Lǐ Shànlán.

9.3 The establishment of new types of school

The first establishment of new-style schools was carried out by missionaries of various denominations. In 1839 Brown established the first church school in Macao. After the Opium Wars the churches moved further into the hinterland and a lot of schools were established that also produced their own text-books. Because of the unequal treaties signed by the Qīng government the various Western countries increased their aggressive activities. Besides increased aggressiveness in military, political, and economic areas, cultural aggression was also greatly strengthened. In the year 1845 the missionaries established what was to become St. John's College in Shànghǎi and in 1864 they established the Language Institute in Shāndōng (山東) province. In 1874 they established Gesu (格致 Gé Zhì) College in Shànghǎi, and in 1888 the Foreign Language College in Beijing. These institutions were among the first-established of the famous church schools.

After the 1911 revolution these sorts of church school still continued to develop, and new schools were established or a number of schools amalgamated to form a bigger school. At the same time they started establishing universities. Later on these church schools became part of the educational system under the half feudal, half colonial rule of the Nationalist government. (See also Latourette 1929/1966, pp. 441–51.)

The earliest new type of school established by the Chinese was the Běijīng 'Foreign Languages Institute'.

The supporters of the Westernization movement had established an influential position in politics after the Opium Wars, and in the year 1862 they established the Foreign Languages Institute. This was a new sort of school modelled on the Western educational system. Initially this school only had the task of teaching foreign languages (English, French, and Russian). In 1866 the Mathematics Department was founded following a proposal to the Emperor by Prince Yí Xīn (奕訢), who was the leader of the Westernization movement. The reasons put forward in the proposal for adding a Mathematics Department were as follows:

Because the foreigners manufacture machinery, arsenals, etc., all of which are based on astronomy and mathematics. At this moment there are negotiations in Shànghǎi and Zhèjiāng (浙江), etc. to buy various types of steamships and other materials, but if we do not start from the very basic foundations what we learn will be just superficial and without practical use. . . . We now propose establishing another department, . . . for the use of logical reasoning, methods of manufacturing and observation, methods of calculation. If we can concentrate on being practical and learn all the essentials, then this is the path to strengthening China.

Of course this sort of argument presented by the foreign affairs party was just a front. There were some arguments even within the ruling class as to whether to carry out 'Westernization', but eventually the Mathematics Department was founded and Lǐ Shànlán was invited to be the chief instructor. According to the Regulations of the Foreign Language Institute the complete course was eight years. Mathematics courses started from the fourth year: in the fourth year there were 'introduction to mathematical theory' and 'algebra', in the fifth year *Euclid's Elements* was studied, 'plane trigonometry' and 'spherical trigonometry' and in the sixth year 'calculus' and 'navigation'. From the start of this educational programme in 1866 up to 1895 AD there was very little change.

Later, following a proposal by the Northern warlord Lǐ Hóngzhāng (李鴻章), an institute was also established in Shànghǎi. This institute was called the Institute of Languages (廣方言館 Guǎng fáng yán guǎn). In a report to the Emperor on the courses in the Institute of Languages by Lǐ Hóngzhāng it says: 'The mathematics classes start after lunch. No matter whether it is calculation with pen and paper or abacus calculation they begin with addition, subtraction, multiplication, and division. In the middle of the course the *Ten Mathematical Manuals* should be learned.'

About the time the Foreign Languages Institute and the Institute of Languages were being established, other places also founded various types of special school and military school. Amongst them the earliest is the Shipbuilding Management College established by Zǔo Zōngtáng (左宗棠) in the Mǎwěi (馬尾) shipyard of Fújiàn (福建) province in 1866. In 1880 the Northern Naval Academy was established in Tiānjiān (天津) city. After 1885 the Tiānjīn Military Academy and the Guǎngdōng (廣東) Army Academy were successively established. Most of these schools had some mathematics courses. For example, the Northern Naval Academy in Tiānjīn had algebra, geometry, plane trigonometry, trigonometric functions, series, mechanics, astronomical calculations, land surveying, and other courses.

In addition, towards the end of the 19th century and at the beginning of the 20th century, various types of college and institute also appeared. For example, the Self-Strengthening College (自强學堂, Zì qiáng) in Húběi (湖北), the College of Húběi and Húnán (兩湖書院 Liǎng hú shū yuàn), the Specialist Institute (精舍, Jīng shè) in Shāndōng, the Wèi Jīng Institute (味經學舍 Wéi jīng xué shè) in Shǎnxī (陝西), and the Real Learning Institute (實學館 , Shí xué guǎn) in Guāngzhōu (廣州). These colleges were modelled on the Běijīng Foreign Languages Institute and most of them had some mathematics courses. Some of these schools were the power bases of the foreign affairs party and the warlords' hierarchy, some were platforms for publicizing the reformists' improvements and for drumming up support for reform.

A reform movement started with the constitutional reform in 1898 AD. After the reformation, the Emperor Guāng Xù (光緒) ordered that high schools, secondary schools, primary schools, voluntary schools, adult schools, and so on, be founded as soon as possible in all the provinces and at the same time the Capital University (that is, the forerunner of the present Běijīng University) was planned. This is generally recognized as the beginning of the establishment of the new university system in China.

Although the Reform Movement of 1898 only lasted for a hundred days before being defeated, nevertheless the movement for establishing new school systems and abolishing the civil service examinations continued to gather support. Universities such as the University of Jìn (晋) province (i.e. Shānxī, 山西 , province), the University of the Two Hú's (Húnán and Húběi), Húnán University, etc. were established successively in other provinces. As far as the regulation of education is concerned there were the so-called 'Imperially decreed Regulations for Schools' of 1902 and also the so-called 'Proposed Regulations' of 1903. In 1905 the civil service examinations were completely abolished and the central government therefore established the 'Education Department' (學部 , Xué bù).

In regard to mathematics syllabi, the regulations in the 'Proposed Regulations' were: for junior primary schools (entry at seven years, five years to complete) and for senior primary schools (four years to complete), arithmetic is studied beginning with 'names for numbers', recording numbers, including the four operations and calculation with pen and paper and abacus, ratios, equalities and also calculations involving measurement, weights, volumes, lines, and money. Secondary schools (five years to complete) study arithmetic, algebra, geometry, trigonometry, etc. Higher institutes (three years) were divided into the three faculties of law, engineering, and medicine with lecture courses on algebra, geometry, analytical geometry, calculus, etc. The Universities were divided into six 'branches' (門 , mén — 'branches' are equivalent to departments in later tertiary institutions); mathematics was one of the branches and the mathematics courses were: differentiation and integration, geometry, algebra, practical calculations, mechanics, number theory, partial differentiation, theory of equations, supplements to algebra and number theory, introduction to theoretical physics and demonstrations for the above courses, experiment in physics, etc.

After the revolution in September 1912, the government announced 'Amendments to the Education System' based on the foundation of the so-called 'Proposed Regulations' of the Qīng Dynasty. These 'amendments' made minor changes in the ages and years at each stage as compared with the old 'Proposals'. However, the 'Proposed Regulations' were not put into effect throughout the country, whereas the 'Amendments to the Education System' were essentially carried out throughout the primary and secondary schools everywhere after the 1911 revolution. As far as the depth and breadth of influence is concerned, that of the 'Amendments to the Education System' greatly exceeded that of the 'Proposed Regulations'. The new education system (including mathematical education) was then established in its initial stages. In the new school system mathematics became a compulsory subject amongst the important subjects that were to be studied by all students.

9.4 The changes in mathematical textbooks

In the process of changing the school system textbooks were also correspondingly changed.

Initially, after all kinds of church schools were being planned throughout

China, there was the translation and editing of some textbooks, including some mathematics books. Before this, as early as the beginning of the second entry of Western mathematics during the 1850s, a set of mathematics books had already been translated by Lǐ Shànlán and others. In the 1870s another set of mathematics texts was also translated by Huà Héngfāng and others. As mentioned above Lǐ Shànlán's 'objective in translating books' was only 'to show off his specialized knowledge' so the books he translated were used by a small number of mathematicians as reference texts and were not suitable for general use as textbooks. When the books translated by Huà Héngfāng were later published by the Jiāng Nán (江南) Manufacturing Company, some schools in various places used them as textbooks. Not long afterwards the Peking Foreign Languages Institute, the Yì Zhì society (益智) and the Guāng Xué Society (廣學) in Shànghǎi, etc. also became involved in translating Western scientific and technical books in order to supply some of the required textbooks.

In the final years of the 19th century and the first few years of the 20th century, the various levels of the Qīng Dynasty schools, private schools, and church schools widely adopted the following mathematics textbooks:

Pen Calculation (筆算數學 , *Bǐ suàn shù xué*), written jointly by the American Presbyterian Missionary Calvin Wilson Mateer and Zōu Liwén (鄒立文). In the text there is the 1892 preface by Mateer and the complete book has 24 chapters. There were two editions: the literary and the modern Chinese.

Outline of Algebra (代數備旨 , *Dài shù bèi zhǐ*) written by the American Mateer, jointly translated by Zōu Liwén and Shēng Fúwéi (生福維), published by the Sino-American Publishing Company (美華 , Měi huà) in 1891, a total of 13 chapters.

Elements of Geometry (形學備旨 , *Xíng xué bèi zhǐ*), written by the American Loomis, jointly translated by the American Mateer, Zōu Liwén and Liù Yǒngxī (劉永錫), published by the Yì Zhì Book Company in 1885, a total of 10 chapters.

Elements of Plane Trigonometry (八線備旨 , *Bā xiàn bèi zhǐ*) written by the American Loomis, freely translated by the American A.P. Parker (Pān Shènwén, 潘慎文) and edited by Xiè Hónglài (謝洪賚), published by the Sino-American Publishing Company in 1894, a total of four chapters.

The Elements of Analytical Geometry (代形合參 , *Dài xíng hé shēn*), written by the American Loomis, freely translated by A.P. Parker edited by Xiè Hónglài, published by the Sino-American Publishing Company in 1893, a total of three chapters (and an appendix).

These texts became very popular in a short time. According to incomplete statistics *Pen Calculation* was reprinted 32 times in the period of 10 years from 1892 to 1902; *Outline of Algebra* was reprinted 10 times between 1891 and 1907; *Elements of Geometry* was reprinted 11 times between 1885 and 1910. Other texts were also reprinted many times. However, the fact that they were reprinted many times does not mean that these books were well edited. The main reason was that because of the changes in the school system at that time and the abolition of the civil service examination system, new-style schools were founded in very many places and a set of textbooks was necessary and in great demand.

Most of these books were printed with lead type or lithographed with diagrams attached. In general, in the equations recorded in these books Arabic numerals, 1, 2, 3, . . . were usually adopted to replace the Chinese numerals , , ,. . . (but note that Lǐ Shànlán used Chinese numerals in writing equations and not Arabic numerals); generally the present-day mathematical symbols $+$, $-$, $\times$, $=$, $>$, $<$, $\sqrt{}$, etc. were also adopted. However, the books still used the Chinese characters 甲, 乙, 丙, 丁, . . . (jiǎ, yǐ, bǐng, dǐng, . . .; the Heavenly Stems) and 子, 丑, 寅, 卯, . . . (zǐ, chǒu, yín, mǎo, . . .; the Earthly Branches), for variables instead of adopting the Western letters a, b, c, d . . . and x, y, etc. These books also followed the traditional column layout in the way the text was printed. Although the form of the book was still conservative, nevertheless the content was essentially the same as that in the mathematics textbooks commonly used in the various countries of the world at that time and therefore was different from in the past. These texts were used in the schools as textbooks for some time and educated the younger generation then. To the present day in China one can

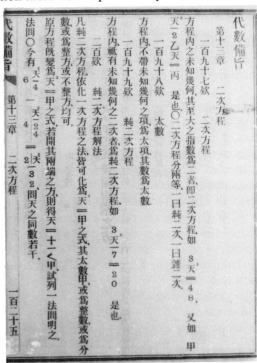

Fig. 9.3 *Outline of Algebra.*
(代數備旨, *Dài shù xué zhǐ*).

still get from the older generation an impression of the circulation of these textbooks then and can still hear how they received their preliminary mathematical education through these textbooks.

It is worth noting that the book *Pen Calculation* was published in another edition in modern Chinese (this is the so-called Mandarin edition) besides an edition in the classical Chinese literary style. This was the first textbook to use the vernacular for writing mathematics.

A significant number of various sorts of solutions to problems, analyses of exercises and similar texts were produced for these texts for convenience in teaching and studying the books mentioned above.

Most of the translating and editing of the mathematics texts in the few decades from Lǐ Shànlán and others' translations to the edition of the *Pen Calculation* was done by foreigners. After the Reform Movement and in particular after the abolition of the civil service examinations in 1905, publishers tried to outdo one another in publishing various sorts of textbooks because of urgent demands from all sorts of schools. Among these publishers were the Shànghǎi Commercial Press, the Scientific Society Publications Institute, the Civilization Book Company, the Yì Zhì Book Company etc. — more than 20 companies. The various publishers started editing and writing various sorts of textbooks, including mathematics texts.

After the set of 'Modern Textbooks' (最新教科書 , Zuì xīn jiào kē shū) was published by the Commercial Press, other textbooks printed by other book companies were gradually supplanted by the Commercial Press books. Before the 1911 revolution the textbooks produced by the Commercial Press were in an absolutely advantageous position. And the most popular mathematical textbooks used were books printed by the Commercial Press.

Among these books, some had already changed to writing mathematical equations along the line (that is, the same as the arrangement of equations in books nowadays), but other descriptive works still used the old column form. As to the contents of these books, they were still the editions and translations of various textbooks used in the countries of Europe, America and Japan.

Because of the change in school systems by the abolition of the civil service examinations, new-style schools were established; through the change in textbooks the old set was replaced by a new one. Besides abacus calculation, which has been preserved and is still widely used today in everyday life, all the rest of the ancient mathematics of China blended into the stream of the development of world mathematics. Around the time of the revolution of 1911 the number of Chinese students going overseas to study mathematics increased steadily. By the 1920s Chinese mathematicians had started to achieve some relatively valuable results in some of the branches of modern mathematics. Chinese mathematics had embarked on the long road of development to a new era — the era of modern mathematics.

Appendix 1: Language

Pronunciation

The *sound* of Chinese characters is indicated in this book using the Pīnyīn (拼音 ; literally sound assembly) system. This means that words can be read almost as in English. The basic exceptions are a few consonants which look strange to Western eyes but are easily remembered:

    c = ts
    q = ch
    x = sh
    zh = j

(otherwise z = dz). For vowels the exceptions are: i after h is pronounced close to the e in 'the', otherwise like ee in 'bee'. e behaves like i when on its own or after h but otherwise is pronounced like e in 'Fred'. ü is pronounced like French u or German ü.

In more detail: each character corresponds to a single sound and is represented by a single syllable. In the middle of the syllable is a group of vowels e.g. a, iao, uai. At the beginning is either one or a pair of consonants or, occasionally, nothing (in which case the word starts with a vowel group beginning with either a or e). Likewise the syllable ends with nothing (after the vowel group), n or ng. (Thus the word Guānguāng consists of the two syllables guān and guāng. If a syllable ending with a vowel is followed by one beginning with a vowel the break between syllables is indicated by '. Thus gǔ'ān consists of gǔ and ān.)

Tones

Each vowel group has four tones of voice in which it can be pronounced and these are indicated by accent (diacritical) marks. These are exemplified by the natural pronunciation of 'home' in the following examples (after Newnham 1971):

'Home, home on the range' (sung)	First tone, written hōme	
'Are you going home?'	Second tone,	hóme
'Surely you can't be going home?'	Third tone,	hǒme
'I'm going home (so there!)'	Fourth tone,	hòme

The pitches of the vowel groups vary thus (after DeFrancis 1963)

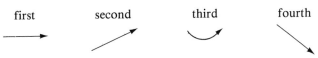

first	second	third	fourth

Further reading

Hawkes (1973, p. 11ff.) for brief but thorough notes; Newnham (1971) for 'an introduction to the Chinese language for non-learners' and DeFrancis (1963, pp. xxii–xxxiv) for the student of the Chinese language.

Writing

Chinese characters are formed from eight basic strokes 、, 一, ｜, 乀, 〆, 亅, 丿, 乀. Just remember 'forever' = yǒng = 永, which uses all the types of stroke. The basic form of Chinese characters has not changed significantly from the time of the Hàn Dynasty (206 BC–220 AD), except for recent simplifications. The strokes have a specific order: basically, top to bottom, left to right, outside before inside (except that the inside is completed before the sealing stroke that completes the outside). See also Newnham (1971) and DeFrancis (1963).

Since the characters for the number 1–9, that is, 一, 二, 三, 四, 五, 六, 七, 八, 九, are so simple, it is not surprising that there is no difference between numerals and number words in ancient Chinese. Likewise it was quite late when symbols were used for + , − , etc. rather than the single characters. Thus there is no difference between literal algebra and symbolic algebra in ancient Chinese mathematics. (See also Hoe 1977, pp. 39ff.)

Radicals

Characters are often composed of two or more simpler characters or units. In addition, similar looking characters often have similar sounds. A similarity may also occur because of a component unit called the 'radical'. The radicals, of which there are at present 226, provide one classification system for the thousands of Chinese characters currently used. (The other common classification systems are by the number and order of the strokes and by the Pīnyīn transcriptions arranged as in an English dictionary. Three thousand characters are enough to read a newspaper but estimates of the total number of extant characters exceed 40 000.)

As an example, the character 訌 (hòng, noisy) has two component units which are both characters: 言 (yán, speech) and 工 (gōng, worker). Its sound is clearly close to that of 工, gōng, but its meaning is close to that of its radical 言, yán.

Appendix 2: Chinese books

Production

Ancient works were originally produced by hand copyists. Paper was invented in China about 100 AD. Wood-block printing (xylography), whereby a whole page was carved at a time, started about 800 AD.

In the process of reproducing texts, glosses (that is, explanatory comments) were added from time to time, usually in smaller characters, just as in the West glosses were added to the Bible, Euclid's Elements, etc. However, revisions did sometimes affect the original text. Nevertheless the ascription of the book's authorship remained unchanged. This often makes it difficult to determine the original date and content of an ancient book.

Titles

Needham (1959) provides a wonderful bibliography of relevant Chinese works. (However he used a slightly modified form of the Wade – Giles transcription of Chinese characters, rather than the Pinyin normally used today.) In order not to introduce yet another set of translations of titles (since there are already too many variants) we have, so far as was possible, retained Needham's translations of titles. This has some disadvantages (cf. Hoe 1977) but the difficulty of translating Chinese titles is that (a) they are usually very terse and (b) they are often ambiguous and play on words (characters) extensively.

We have made one major and one minor exception to the above. The *Zhōubi suànjīng* has been referred to only by this title. 'Zhōu (周)' is the name of an ancient Chinese dynasty — the one when the book was written. But the very same character also means 'perimeter' or 'circular path'. Needham (1959) translates *Zhōubi suànjīng* therefore as the *Arithmetical Classic* (suànjīng) *of the Gnomon* (bi) *and the Circular Paths* (zhōu) *of Heaven*. This certainly gives a good idea of the content as the book is indeed about astronomy. Another possible title would be the 'Zhōu Gnomon Classic of Calculation', since suàn (算) means calculation; but there are many other possibilities, none as succinct as *Zhōubi suànjīng*. The minor exception is that we have replaced Needham's *Shou Shih Calendar* by the literal translation *Works and Days Calendar*.

Measures

In ancient times Chinese books were carved on bamboo strips joined into rolls. Such a roll was called a juàn (卷). Sets of rolls were stored in boxes. Later, silk and paper scrolls were used. On one scroll there could be several short works but equally a large work could occupy several scrolls. It is therefore hard to translate these measures uniformly into the English 'chapter', 'book', 'volume' and we have tried to give an indication of the size of works by choosing the closest English equivalent for the particular context.

269

Appendix 3: Chronology

Xià Dynasty		About 21st to 16th century BC	
Shāng Dynasty		About 16th to 11th century BC	Oracle-bone numerals
Zhōu Dynasty	Western Zhōu Dynasty	11th century to 771 BC	Bronze script numerals
	Eastern Zhōu Dynasty	770–256 BC	
	Spring and Autumn Period	770–476 BC	Decimal numeration Row and column forms
	Warring States	475–221 BC	
Qín Dynasty		221–201 BC	
Hàn Dynasty	Western Hàn	206BC–24 AD	*Zhōubì suànjīng*: the oldest of the mathematical classics. Gōugǔ (= Pythagoras' theorem)
	Eastern Hàn	25–220 AD	*Nine Chapters on the Mathematical Art*: about first century AD, the most important of all ancient Chinese mathematical books
Three Kingdoms	Wèi	220–265 AD	
	Shǔ Hàn	221–263 AD	
	Wú	222–280 AD	
Jìn Dynasty	Western Jìn Dynasty	265–316 AD	Liú Huī, *Commentary on the Nine Chapters* and *Sea Island Mathematical Manual*
	Eastern Jìn Dynasty	317–420 AD	
Northern and Southern Dynasties	Southern Dynasty		
	Sòng	420–479 AD	Zǔ Chōngzhī,
	Qí	479–502 AD	evaluation of π
	Liáng	502–557 AD	Mathematical books:
	Chén	557–589 AD	*Master Sūn's Mathematical Manual*,
	Northern Dynasty		

	Northern Wèi	386–534 AD	*Mathematical Manual of the*
	Eastern Wèi	534–550 AD	*Five Government*
	Northern Qì	550–577 AD	*Departments,*
	Western Wèi	535–556 AD	*Xià Hòuyáng's*
	Northern Zhōu	557–581 AD	*Mathematical Manual, Zhāng Qiūjiàn's Mathematical Manual.* Chinese remainder theorem
Sui Dynasty		581–618 AD	
Táng Dynasty		618–907 AD	*Ten Mathematical Manuals*
Five Dynasties	Later Liáng	907–923 AD	
	Later Táng	923–936 AD	
	Later Jìn	936–946 AD	
	Later Hàn	947–950 AD	
	Later Zhōu	951–960 AD	
Sòng Dynasty	Northern Sòng	960–1279 AD	Zenith of mathematics in ancient China, Lǐ Zhì
	Southern Sòng	1227–1279 AD	(Lǐ Yě), Qín Jiǔsháo, Yáng Huī and Zhū Shìjié.
Liáo Dynasty		916–1125 AD	'Method of the four unknowns'
Xià Dynasty		1032–1227 AD	
Jìn Dynasty		1115–1234 AD	
Yuán Dynasty		1271–1368 AD	
Míng Dynasty		1368–1644 AD	
Qīng Dynasty		1644–1911 AD	
Republic of China		1912–1949 AD	
People's Republic of China		1949 AD	

Appendix 4: History

China has long been almost cut off by land from the West by the loftiest mountain ranges and formidable deserts. Sea travel has not compensated for this. It is not surprising, therefore, that Chinese culture developed practically independently of Europe's. What is suprising, and not sufficiently well-known in the West, is the extent and age of that culture.

It is only since the 1949 revolution that the first concrete evidence from the previously legendary Xià (夏) Dynasty (c. 21st to 16th centuries BC) has been excavated. From the Xià to 1911 China was usually ruled by a royal house or dynasty. The history of China is classified by these dynasties and within these dynasties by the names of the various reigns.

The next dynasty after the Xià was the Shāng (商) Dynasty (c. 16th to 11th centuries BC), which was also thought until recently to be legendary, but which is now significantly authenticated (see e.g. Keighley 1978; Ho Ping-Ti 1975). In this period basic mathematics can be discerned (see Chapter 1). Chinese writing from this period is not significantly different from present-day Chinese writing. In the Shāng dynasty China was quite small and centered round Ānyáng (安陽). Succeeding dynasties came about because of either the decline of the previous one or its being conquered by a neighbouring court. However there were several periods of instability: for example the Spring and Autumn period (770–475 BC), so-named from a book called the *Spring and Autumn Annals* (春秋 , *Chūn qiū*).

Before the end of the next dynasty, [the Zhōu (周)] in 221 BC, the basic classical books had already been written. For example, Confucius (Kǒng Fūzi, 孔夫子 , 551–479 BC) or his immediate disciples wrote in that period.

In 221 BC the first Qín (秦) emperor unified China for the first time. He was the first to establish uniform weights and measures. He also standardized the writing of Chinese characters. In 213 BC however he ordered the 'burning of the books'. Fortunately copies of some of these books have subsequently been found to have survived. In the Qín Dynasty (221–201 BC) the Great Wall of China was completed.

The Hàn (漢) Dynasty (206 BC–220 AD) commanded a great area of central Asia and developed the silk trade with Europe. This was a time of great prosperity and of great cultural growth. The *Nine Chapters on the Mathematical Art* (see Chapter 2) was written by this time. The *Chronicles* of Sīmǎ Qiān (司馬遷 , see Chavannes 1895) were written under the Hàn Dynasty. Chinese characters have hardly changed from that time until this century. The Hàn Dynasty was also the first to establish a centralized government and a civil service to cope with a population of sixty million people (in 2 AD). Taxes were either monetary or extracted in the form of *corvée* labour. This corvée labour meant each farmer had to spend about one month per year working on roads, canals, palaces, etc. for the government plus, sometimes, military service.

After the Hàn there was a period of disunity that lasted until the Suí (隋) Dynasty (581–618 AD). During the period of disunity, Buddhism, which had entered China in the first century AD, developed strongly. Buddhism was at its height in the fourth to eighth centuries AD.

Although the Sui dynasty lasted only 37 years it was followed immediately by the Táng (唐, 618–907 AD) which produced another era of outstanding commercial and cultural growth. The expansion of Táng was halted by Islamic conquest from the West.

The Táng was again followed by a period of disunity — the Five Dynasties (907–960 AD) — but the arts flourished again under the Sòng (宋) Dynasty (960–1279 AD). The Sòng was overthrown by the Mongols who invaded from the North under Chinggis Khan (Ghengis Khan). They formed the Yuán (元) Dynasty (1271–1368 AD). Marco Polo stayed in the court of Kublai Khan for seventeen years during this dynasty. He was the most famous of a number of Western travellers who have left well-documented accounts of their journeys (see Sykes 1933 and Hudson 1931).

Despite the troubles of the Mongol invasion Chinese science prospered during this time and the zenith of ancient Chinese mathematics was reached.

The gradual weakening of Mongol rule encouraged many rebels and eventually one of these, Zhū Yuánzhāng (朱元章) founded the Míng (明) Dynasty (1368–1644 AD). During the reign of this dynasty Western missionaries gained admittance to the Chinese mainland, starting with Matteo Ricci. Information about China began to be eagerly sought in the West. However the Manchus invaded from the North and eventually founded the Qīng (清) Dynasty. This lasted from the end of Míng until 1911 AD, despite invasions by foreign powers. China then became dominated by foreign powers until the revolution of 1949. Since then China has again become a truly independent nation.

Further reading

For the period covered by this book Reischauer and Fairbank (1960) is a superb book and clearly the best available. A shortened version is in Craig, Reischauer, and Fairbank (1973). An excellent recent history of China along with an emphasis on very early Chinese history is Milston (1978). An up-to-date encyclopaedia that provides basic information is Hook (1982).

Bibliography

Adamo, M. (1968). La matematica nell'antica Cina, *Osiris* **15** (1968), 175–95.

Ang Tian-Se (1976). The use of interpolation techniques in Chinese calendar, *Oriens Extremus*, Jahrgang **23** (1976), 135–53.

—— (1977). Chinese computation with the counting-rods, *Kertas-Kertas Pengajian Tionghua* (Papers on Chinese Studies) I (1977), 97–109.

—— (1978). Chinese interest in right-angled triangles, *Hist. Math.* **5** (1978), 253–66.

—— and Swetz, F.J. (1984). A brief chronological and bibliographic guide to the History of Chinese Mathematics. *Hist. Math.* **11** (1984), 39–56.

Aristotle (1908/1952). Works (1908, 2nd edn. 1952, ed. & c. W. D. Ross). Clarendon Press, Oxford.

Bennett, A. A. (1967). *John Fryer: The introduction of western science and technology into nineteenth century China.* East Asian Research Center, Cambridge, Mass.

Berezkina, E. J. (1957). Drevnekitajski trakat 'Matematika v devjati knigach', *Istoriko-Matematičeskie Issledovanija* **10** (1957), 423–584.

Bézout, E. (1764). Recherches sur le degré des équations résultantes de l'évanouissement des inconnues. *Mémoires de l'Académie royale des Sciences* (1764), 288–338.

Biernatzki, K. L. (1855/1973). *Die Arithmetik der Chinesen.* Reprinted Dr. Martin Sändig, Walluf bei Wiesbaden (1973) from *Crelle's Journal* f.d.M.Bd.LII, Heft 1 (1855).

Biot, E. (1841) Traduction et examen d'un ancien ouvrage chinois intitulé: TCHEOU-PEI littéralement: 'Style ou signal dans une circonference'. *J Asiat.* **11** (1841), 593–639 (actual translation, 599–639). Note supplémentaire *J. Asiat.* **13**, (1842), 198–202.

Bombelli, R. (1572/1966). *L'Algebra* (three books), Bologna (1572). Reprinted with new title page only, 1579. Complete edition (Books I–V, ed. E. Bortolotti), Feltrinelli, Milan (1966).

Chang Kwang-chih (1977). *The archaeology of ancient China*, (3rd edn). Yale University Press, New Haven and London.

—— (1980). *Shang civilization.* Yale University Press, New Haven and London.

Chavannes, E. (1895). *Sze-ma Ts'een, Les Memoires Historiques*; traduites et annotées par E. Chavannes (3 Vol.). Paris.

Colebrooke, H. T. (1817/1973). *Algebra, with arithmetic and mensuration from the Sanscrit of Brahmegupta and Bháscara*, Murray, London (1817). Reprinted Dr Martin Sändig, Walluf bei Wiesbaden (1973).

Couling, S. (1917/64). *The Encyclopaedia Sinica.* Kelly and Walsh, Shanghai (1917). (Literature House 1964).

Craig, A. M., Reischauer, E. O., and Fairbank, J. K. (1973). *East Asia: tradition and transformation.* Houghton Mifflin, Boston.

De Francis, J. (1963). *Beginning Chinese*, Yale University Press, New Haven.

De Morgan, A. (1835). *Elements of algebra preliminary to the differential calculus* . . . John Taylor, London. (2nd edn, London 1837.)

—— (1838). *An essay on probabilities* . . . London.

Dunne, G. H. (1962). *Generation of giants*, University of Notre Dame Press, Notre Dame, Indiana, USA.

Gillispie, C. C. (1970–76). *Dictionary of scientific biography*, Scribner, New York.

Gillon, B. S. (1977). Introduction, Translation, and Discussion of Chao Chun-Ch'ing's 'Notes to the diagrams of short legs and long legs and of circles and squares', *Hist. Math.* **4** (1977), 253–93.

Hawkes, D. (1973). Translation of *Cao Xueqin: The story of the stone*, Penguin Books, Harmondsworth, England.

Herschel, Sir J. F. W. (1849). *Outlines of astronomy*, Longmans, London.

Ho Peng Yoke (1965). The lost problems of the Chang Ch'iu-chien Suan Ching, a fifth-century Chinese mathematical manual, *Oriens extremus* Jahrgang 12 Heft 1 (1965), 37–53.

—— (1970a). Ch'in Chiu-Shao in Gillispie (1970), Vol. III, 249–56.

—— (1970b). Chu Shih-Chieh in Gillispie (1970), Vol. III, 265–71.

—— (1970c). Li Chih, also called Li Yeh, in Gillispie (1970), Vol. VIII, 313–20.

—— (1970d). Liu Hui in Gillispie (1970) Vol. VIII, 418–25.

—— (1970e). Yang Hui in Gillispie (1970), Vol. XIV, 538–46.

—— (1973). Magic squares in east and west. *Papers on Far Eastern History*, Australian National University, Dept. of Far Eastern History 8 (1973), 115–41.

Ho Ping-Ti (1975). *The cradle of the East*. Chinese University Press, Hong Kong.

Hoe, Jock (1977). *Les Systèmes d'Équations Polynômes dans le Siyuan Yujian (1303)* Collège de France, Institut des Hautes Études Chinoises (1977). [The author's first name is incorrectly printed as 'John'.]

—— (1978). The Jade Mirror of the four unknowns — Some reflections. *Math. Chronicle* **7** (1978), 125–56.

Hook, B. (ed.) (1982). *The Cambridge Encyclopaedia of China*, Cambridge University Press.

Horner, W. G. (1819). A new method of solving numerical equations of all orders by continuous approximation, *Phil. Trans. of the Roy. Soc.* **109** (1819), 308–35.

Hudson, G.F. (1931). *Europe and China: a survey of their relations from the earliest times to 1800*, Arnold, London.

Hymers, J. (1837). *A treatise on plane trignometry*, For J. & J. Deighton, Cambridge (1837), Several edns.

Keightley, D. N. (1978). *Sources of Shang history: the oracle-bone inscriptions of bronze age China*, University of California Press, Berkeley.

Kiang, T. (1972). An old Chinese way of finding the volume of a sphere. *Math. Gazette* **56** (1972), 88–91.

Kobori, Akira (1970). Tsu Ch'ung-Chih in Gillispie (1970), Vol. XIII, 484–5.

Lam Lay-Yong (1966). On the Chinese Origin of the Galley Method of Arthimetical Division. *British Journal for History of Science* **3** (1966), 66–9.

—— (1969). On the Existing Fragments of Yang Hui's Hsiang Chieh Suan Fa. *Archive for History of Exact Sciences* **5** (1969), 82–6.

—— (1969a). The geometrical basis of the Ancient Chinese square-root method. *Isis* **61** (1969), 96–102.

—— (1972). The *Jih-yung suan fa*: An elementary arithmetic textbook of the thirteenth century. *Isis* **63** (1972), 370–83.

—— (1977). *A critical study of the Yang Hui Suan Fa, a thirteenth century Chinese mathematical treatise*. Singapore University Press.

—— (1979). Chu Shih-chieh's Suan hsüeh ch'i-meng (Introduction to Mathematical Studies). *Archive for History of Exact Sciences* **21** (1979), 1–31.

—— (1980). The Chinese connection between the Pascal triangle and the solution of numerical equations of any degree. *Hist. Math.* **7** (1980), 407–24.

—— (1982?). The historical development of polynormial equations in China, *Proc. 1st International Conference on the History of Chinese Science*, Leuven 1982 (to appear).

—— and Ang Tian-Se (1984). Li Ye and his *Yi Gu Yan Duan* (Old Mathematics in Expanded Sections). *Archive for History of Exact Sciences* **29** (1984), 237–66.

Legge, J. (tr.) (1882/1966). *The sacred books of China Pt. II the Yi King* [Yi Jing], Vol. XVI translated by J. Legge ed. M. Muller. Clarendon Press, Oxford (1862). Reprinted Motilal Banarsidass, Delhi (1966).

—— (1891/1966). *The sacred books of China Pt. I*, Taoist books, Vol. XXXIX Clarendon Press, Oxford (1891). Reprinted Motilal Banarsidass, Delhi (1966).

Libbrecht, U. (1973). *Chinese mathematics in the thirteenth century, the Shu-shu chiu-chang of Ch'in Chiu-shao*. MIT Press, Cambridge, Mass. USA.

Loomis, E. (1843). *Elements of geometry and conic sections*. Harper & brothers, New York. [At least 25 edns!]

—— (1848). *Elements of plane and spherical trignometry*. Harper & brothers, New York. [At least 55 edns!]

—— (1879). *The elements of analytical geometry*. Harper & brothers, New York.

Latourette, K. S. (1929/1966). *A history of Christian missions in China*. S.P.C.K., London (1929). Reprinted, Ch'eng-Wen, Taipei (1966).

Martzloff, J-C. (1981). *Recherches sur l'oeuvre mathématique de Mei Wending (1633–1721)*. College de France, Institut des Hautes Etudes Chinoises.

Mikami, Y. (1913, 1974). *The development of mathematics in China and Japan*. Leipzig (1913). 2nd edn Chelsea, New York (1974).

—— (1928). The Ch'ou-Jen Chuan of Yuan Yuan, *Isis* **11** (1928), 123–6.

Milston, G. (1978). *A short history of China*. Cassell, Stanmore, NSW.

Mungello, D. E. (1977). *Leibniz and Confucianism: the search for accord*, University Press of Hawaii, Honolulu.

Napier, J. (1614). *Mirifici logarithmorum canonis descriptio, ejusque usus*, . . . Andrew Hart, Edinburgh (1614).

Needham, Sir J. (1959). *Science and civilisation in China*, Volume 3, part I. Cambridge University Press, Cambridge.

Newnham, R. (1971). *About Chinese*. Penguin Books, Hardmondsworth, England.

Reischauer, E. O. and Fairbank, J. K. (1960). *East Asia, the great tradition*. Allen & Unwin, London.

Ricci, M. (1953). *China in the sixteenth century: the journals of Matthew Ricci: 1583–1610*. Tr. L. J. Gallagher, Random House, New York.

Rowbotham, A. H. (1966). *Missionary and Mandarin*. Russell & Russell, New York.

Ruffini, P (1804). *Sopra la determinazione delle radici delle equazioni numeriche di qualunque grado* Modena (1804). Reprinted in Vol. 2, *Opere Mathematichi di Paulo Ruffini* ed. E. Bortolotti. Editioni Cremonese, Rome (1953), 281–404 and note 50, p. 506.

Sarton, G. (1947). *Introduction to the history of science*, Vol. III, Science and Learning in the Fourteenth Century. Williams & Wilkins, Baltimore.

Sykes, Sir P. (1933). *A history of exploration*. Routledge, London.

Tsien [Ch'ien] Tsun-hsün. *Written on bamboo and silk*. University of Chicago Press, Chicago.

van Hée, L. (1920). Le *Hai Tao Suan Ching* de Lieou. *T'oung Pao* **20** (1920), 51–60.

—— (1926). The Ch'ou Jen Chuan of Yuän Yuän. *Isis* **8** (1926), 103–18.

—— (1932a). Le Classique de l'Ile Maritime, ouvrage chinoise du IIIe Siècle, *Quellen and studien zur Geschichte der Mathematik* (Abteilung B, Astronomie und Physik) 2 (1932), 255–280 (actual translation 269–80).

—— (1932b) Le précieux miroir des quatre éléments. *Asia Major* 7 (1932), 242–70.

Väth, A. (1933). Johann Adam Schall von Bell, S.J., *Veröffentlichungen des Rheinisches Museums in Koln*, Bd. 2. Cologne.

Vogel, K. (1968). *Neun Bücher Arithmetischer Technik*, Ubersetzt und erläutert von K. Vogel. Vieweg & Sohn, Braunschweig.

Wagner, D. B. (1978). Doubts concerning the attribution of Liu Hui's commentary on the Chiu-Chang Suan-shu. *Acta Orientalia* **39** (1978), 199–212.

—— (1979). An early Chinese derivation of the volume of a pyramid: Liu Hui, Third Century AD. *Hist. Math.* **6** (1979), 164–88.

Wang Ling and Needham, J. (1955). Horner's method in Chinese mathematics; its origins in the root-extraction procedures of the Han dynasty. *T'oung Pao* **43** (1955), 345–401.

Wrench, J. W., Jr. (1960). The evolution of extended decimal approximation to π, *Mathematics Teacher* **53** (1960), 644–50.

Zhou Shide (1983). Various types of Carriages; the South Pointing Carriage and Odometer in *Ancient China's Technology and Science*. Compiled by the Institute of the History of Natural Sciences, Chinese Academy of Sciences, Foreign Languages Press, Beijing (1983), 429–35.

Index